about mathematics

prentice-hall, inc., englewood cliffs, new jersey

about

mathematics

richard s. hall

chairman, department of mathematics
willamette university
salem, oregon

Library of Congress Cataloging in Publication Data

HALL, RICHARD S.
About mathematics.

Bibliography: p.
1. Mathematics—1961- I. Title.
QA39.2.H3 510 72-14115
ISBN 0-13-000752-8

about mathematics

by richard s. hall

10 9 8 7 6 5 4 3 2 1

Printed in the United States of America

PRENTICE-HALL INTERNATIONAL, INC., London
PRENTICE-HALL OF AUSTRALIA, PTY. LTD., Sydney
PRENTICE-HALL OF CANADA, LTD., Toronto
PRENTICE-HALL OF INDIA PRIVATE LIMITED, New Delhi
PRENTICE-HALL OF JAPAN, INC., Tokyo

to gretchen
without whose assistance and
encouragement this book would
never have been finished.

contents

preface

This book began as a set of notes for a course offered to liberal arts students at Syracuse University. The majority of these students were freshmen and most of them had decided that their principal interests were not in the sciences. The purpose of the course was to provide these students with a better understanding of the nature of mathematics.

The person whose contact with mathematics has been limited to the high-school courses in geometry and algebra has usually received an erroneous impression about what mathematics is and what mathematicians do. The geometry that he learned, for the most part, was known and compiled by the time of Euclid in 300 B.C. Although the notation used is more recent, the algebraic techniques that he has seen are almost as old. He has little or no knowledge of the mathematics developed in the last 300 years. That such a person should find it difficult to understand how anyone could enjoy or appreciate mathematics should not be surprising; most mathematicians find very little to appreciate in high-school mathematics.

Those students who take "serious" courses in mathematics usually find themselves in a similar position. The courses that they take offer them thorough discussions of very limited topics in mathematics. Seldom, if ever, are they provided with any significant amount of background or a framework for the mathematics they have learned.

It is hoped that this book will be of interest and value to all those who wish to know more *about* mathematics. The mathematical prerequisites for its use are found in the traditional high-school algebra and geometry courses. The reader must also have the perseverance, however motivated, to carefully consider a number of new, often nontrivial, concepts.

The first part of the book is devoted to a historical introduction that considers the mathematics known in the ancient civilizations and the contributions of the Greeks. Each of the other four parts of the book contains a more or less chronologically ordered discussion of the development of ideas and techniques that are important in modern mathematics. The emphasis throughout is on concepts rather than on computational agility or logical rigor. Each part of the book begins with topics that should be recognizable as mathematics to the student and traces the growth and shift of emphasis within a major branch of mathematics. The goal is to explain, by showing the actual historical stages of the metamorphosis, what mathematics is and why it has become the way it is.

There are exercises and discussion questions at the end of each chapter. The exercises are routine and generally phrased so that the answer is provided. They are designed to allow an easy check on the more computational aspects of the material. The discussion questions, on the other hand, seldom have brief or even clear-cut answers. They are meant to be thought about and discussed. The discussions can proceed at a variety of levels depending on the mathematical sophistication of the participants. They may also be used as the starting points for "independent research" by students.

The table of contents should provide those with some mathematical background with sufficient information about the selection of topics. These topics are, in general, simply surveyed and the basic concepts introduced. If each were treated vigorously or in detail, there would be material sufficient for several courses. By proceeding at a fairly rapid pace and resisting the temptation to elaborate further, the material can be covered in a one-semester course by students of average ability and background.

A two-semester course for those less well prepared is possible by proceeding at a slower pace and making judicious use of the problems. Students who have had several previous mathematics courses can complete the entire book in one semester with ample time for discussion of all the problems and the numerous related questions that will arise naturally. (The instructor will undoubtedly find the experience stimulating.)

I wish to acknowledge the assistance and cooperation given me by the staff of the George Arents Research Library and Mrs. Nancy Rude of the Mathematics Library at Syracuse University. I wish also to thank my colleagues William Groening and J. Kevin Doyle for the suggestions they made on the basis of their use of the manuscript of the book.

R. S. H.

about mathematics

part one

introduction

one

mathematics in ancient civilizations

Modern mathematics is the result of centuries of development and refinement. Each succeeding generation of mathematicians has built upon the work of those that preceded. As a result it is necessary to know something of the history of mathematics when trying to understand the nature of the mathematics of today.

No one knows, of course, when man first began to do mathematics. As early as 30,000 B.C. our ancestors carved symbols on animal bones to keep a count of the progress of the phases of the moon. A carved bone with markings that may be a primitive multiplication table has been found near Lake Edward in central Africa and probably dates back to 6000 B.C.

Later civilizations have left massive evidence of the ability to make extensive and accurate measurements. The Great Pyramid at Giza, for example, was built about 2600 B.C. and the errors made in laying out the sides and angles of its square base are extremely small. The famous stone ruins at Stonehenge in England are apparently the remains of a structure built about the same time. The ruins consist of giant stones that have been accurately positioned in large circles. It has been shown recently that the number and location of the stones is such that they can be, and may have been originally, used to predict the movements of the sun and moon in the sky and thereby predict the changes in the seasons and the occurrences of eclipses.[1]

In general, little tangible evidence of mathematical activity has survived from civilizations earlier than those that flourished in Egypt and

[1] For a discussion of one of the recent theories, see Gerald S. Hawkins, *Stonehenge Decoded* (New York: Dell Publishing Co., 1965).

Mesopotamia beginning about 2000 B.C. The sophistication of the mathematics done at that time indicates, however, that man had been doing simple arithmetic and practical geometry for generations. It is only because of the climate in these two regions and because durable materials were chosen on which to inscribe records that we have a considerable amount of information about the mathematical abilities and accomplishments of the Egyptians and the Babylonians.

Egyptian Mathematics

The most extensive record of Egyptian mathematics known today is the papyrus found by A. H. Rhind in 1858. This papyrus is a copy made about 1650 B.C. of an older work that has not survived. The prefatory remarks in the text state that the content has to do with "accurate reckoning" and that "mysterious secrets" are to be revealed.[2] The text of the papyrus is written in the hieratic, or sacred, script rather than the demotic script used for everyday matters. It is reasonable to infer from this that the facts contained in the papyrus were known only to a select group of people.

It was possible to translate the Rhind papyrus almost immediately because of the knowledge gained from the Rosetta Stone which was found by officers of Napoleon's army in 1799. This stone contains an inscription carved in triplicate in Greek, hieroglyphic, and demotic. (Hieroglyphic is the form of hieratic used when carving on stone.) The Rhind papyrus was found to be a textbook for government officials and consists of a large number of practical problems. The solution was provided for each problem. The reader of the papyrus was expected to be able to add and subtract positive whole numbers. The details of multiplying two numbers were shown each time it was done.

Multiplication was accomplished by repeatedly doubling one of the numbers and then adding the appropriate "doubles" to form the product. For example, the products 6 · 13 and 11 · 13 would have been found as shown below.

1	13			\1	13	
\2	26			\2	26	
\4	52	Total 78		4	52	
				\8	104	Total 143

[2] A. B. Chace, L. S. Bull, H. P. Manning, and R. C. Archibald, *The Rhind Papyrus* (Oberlin, Ohio: M.A.A., 1927–1929), two volumes.

Division was done by considering it as the inverse of multiplication. A problem such as "What is the result of dividing 7 by 3?" was solved by answering the question, "What number multiplied by 3 gives 7?". The computation could then be

1	3	
\\$\frac{1}{3}$	1	
\\2	6	Total $2\frac{1}{3}$

The Rhind papyrus contains examples that showed officials how to make the computations necessary in their jobs. There are problems solved showing how to calculate the wages for a group of men, the amount of feed for a number of livestock, the area of a field, and the volume of a granary. The two problems that follow are typical. (Further examples are included in the exercises.)

Problem: One hundred loaves of bread are to be divided among 10 men, including a boatman, a foreman, and a doorkeeper, each of whom is to receive a double share. What is the share of each?

Problem: Find the volume of a cylindrical granary of diameter 9 and height 10.

The answer to the second problem is given in the papyrus as 640 because the Egyptians found the area of circles by a procedure equivalent to the formula $A = (\frac{8}{9}d)^2$. They apparently regarded this formula as exact and used it for all circles. There is no evidence, however, that they had deduced that the ratio of the area of a circle to the square of its diameter is constant (namely, $\pi/4$) and that this same constant can be used to relate the circumference to the diameter ($C = \pi d$).

Egyptian mathematics of this period consisted of a set of procedures that could be used to solve elementary problems concerning simple geometric objects or arithmetic conditions.

Babylonian Mathematics

By about 3000 B.C. the Sumerians had developed an extensive civilization in the area between the Tigris and Euphrates Rivers in what is now Iraq. The Sumerians invented the cuneiform script. They made permanent records by impressing these wedge-shaped marks in soft clay and then baking the clay. By 1700 B.C. the Sumerians had been absorbed by their northern neighbors the Akkadians and the region was ruled by Hammurabi. The political center of this civilization was Babylon.

Thousands of cuneiform tablets made by the Babylonians have survived. The decipherment of these tablets has proved to be very difficult because of the variety of dialects used and the differences caused by the natural modifications of the language over the 2000 years in which the tablets were written. A huge relief carving found about 1850 on a cliff near Behistun, Iran, provided part of the key to translation of the texts. It depicts and gives a multilingual account of the accomplishments of Darius I, king of Persia. Most of the information available about the mathematical content of the tablets is due to the work of Otto Neugebauer since 1930.[3]

The Babylonians inscribed many mathematical tables and problem texts and probably stored them in libraries or archives. The Sumerians used such tables to help them multiply. The later Babylonians used tables of reciprocals, squares and square roots, and cube roots. These tables were probably used not only for mathematical computations but also for the astronomical calculations necessary to keep accurate calendars.

n	n^2	$1/n$	n	n^2	$1/n$
2	4	.30	10	1,40	.6
3	9	.20	12	2,24	.5
4	16	.15	15	3,45	.4
5	25	.12	16	4,16	.3,45
6	36	.10	18	5,24	.3,20
8	1,4	.7,30	20	6,40	.3
9	1,21	.6,40	24	9,36	.2,30

The table above shows the Babylonian values for the squares and reciprocals of some small numbers. The entries in the table are written using the modern symbols for numbers, but many of the entries appear strange because the Babylonian method of grouping numbers into blocks that contain 60 or a power of 60 has been retained. Thus the table shows that

$$8^2 = 1 \cdot (60) + 4 \qquad 9^2 = 1 \cdot (60) + 21 \qquad (12)^2 = 2 \cdot (60) + 24$$

$$\frac{1}{8} = \frac{7}{60} + \frac{30}{(60)^2} \qquad \frac{1}{9} = \frac{6}{60} + \frac{40}{(60)^2} \qquad \frac{1}{12} = \frac{5}{60}$$

[3] For example, see O. Neugebauer and A. Sachs, *Mathematical Cuneiform Texts* (New Haven, Conn.: American Oriental Society, 1945).

Although the modern notation for numbers is decimal, that is, numbers are written in terms of multiples of 10, 10^2, 10^3, . . . , the Babylonian system continues to exist in our units of measurement for time and angles. We have 60 seconds in a minute and 60 minutes in both an hour and a degree of angle measurement.

The Babylonian problem texts contain examples similar to those found on Egyptian papyri. The examples show how to calculate simple areas and volumes and how to find numbers that satisfy given conditions. The circumference of a circle is commonly assumed to be three times the diameter and the area is $\frac{1}{12}$ the square of the circumference. Recently discovered tablets indicate that the Babylonians knew these formulae were not exact and that when more accuracy was desired they divided by the factor .57,36. That is, they were approximating π by the value

$$\frac{3}{.57,36} = 3 \cdot (1.2,30) = 3.7,30$$

which is $3\frac{1}{8}$.

The problem texts show that the Babylonians were much more adept at solving equations than the Egyptians. Their methods will be discussed in the section on algebra.

Summary

The Egyptians and Babylonians were reasonably proficient at elementary arithmetic, geometry, and algebra. They had techniques for solving most of the problems that arose in their daily lives. In fact, the most significant characteristic of the mathematics done at this time is its immediate usefulness. The Egyptians and the Babylonians were not interested in results that did not have practical applications, in general methods for solving problems, or in demonstrating the validity of the procedures they used.

EXERCISES[4]

1. Do the exercise from the Rhind papyrus about paying men with loaves of bread. (Use any method you wish but notice, at least, that it is fairly difficult using Egyptian methods.)

2. (From the Rhind papyrus) One hundred loaves of bread each containing .1 hekat of grain are to be traded for a quantity of beer, each des of which contains .5 hekat of grain. How many des of beer will there be?

[4] The cuneiform problems are adapted from tablets discussed by Neugebauer and Sachs. The papyrus problems are discussed in the volumes by Chace, *et al.*

3. Verify the computations in the following problem from a cuneiform problem text. "An excavation is to be dug in 1 day. The length is 5, the width is 1.30, the depth is 6. The work output of each man is .10 and each man is to be paid .2 pieces of silver a day. What is the area, the volume, the number of workers, and the total wages paid? When you multiply the length and the width, you find the area 7.30. Multiply 7.30 by the depth and find the volume 45. Take the reciprocal of the output and you find 6. Multiply by 45 and find the number of workers 4, 30. Multiply the workers by the wages and find expenses 9. Such is the procedure."

4. What is the value used for π by the Egyptians? How good are the approximations for π used by the Egyptians and the Babylonians?

DISCUSSION QUESTIONS

1. The following is a free translation and simplification of one section of the Rhind papyrus. Deduce as much as you can from this short sample about the mathematics known to the Egyptians of the time it was written. Be as specific as possible, but do not hesitate to use your imagination and make reasonable conjectures from the evidence available.

1	8
2	16
4	32
\8	64

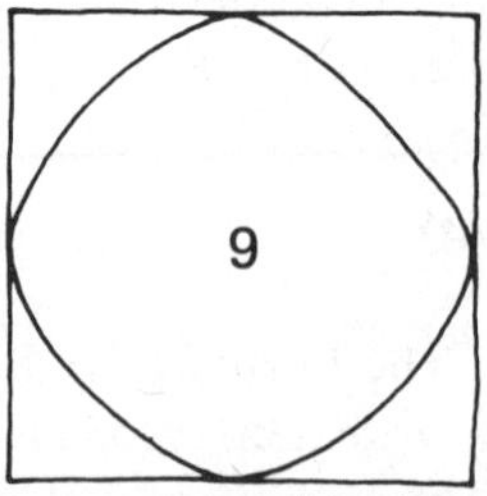

\1	9
2	18
4	36
\8	72
Total	81

Example of a round field of diameter 18 khet. What is its area? Take away $\dot{9}$ of it, namely 2, leaving 16. Multiply 16 by 16: 256. The area is 256 setat. The work is

1	18	1	16
		2	32
$\dot{9}$	2	4	64
		8	128
		\16	256

Example of a triangle of land. What is the area of a triangle of side 10 khet and of base 4 khet? Do it thus:

1	4	1	10
		\2	20
$\dot{2}$	2		

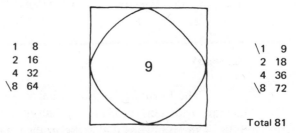

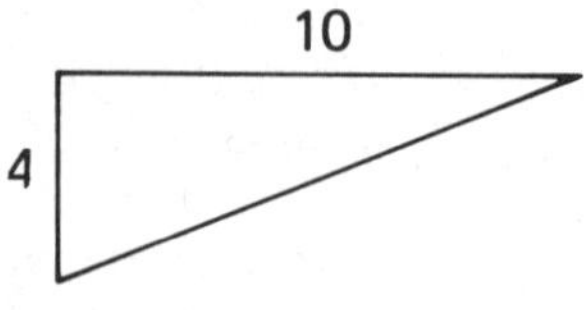

The area is 20 setat.

Example of a cutoff triangle of land. What is the area of a cutoff triangle of land with side 20 khet base 6 khet, cutoff side 4 khet? Add the base to the cut-off side: 10. Take 2 of 10: 5. Multiply 20 by 5. The answer is 100 setat The work is

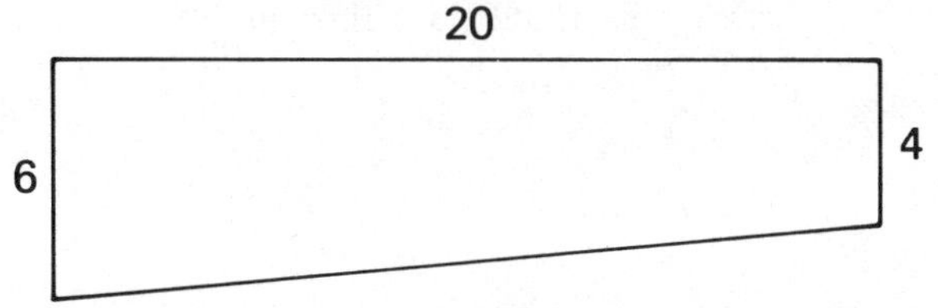

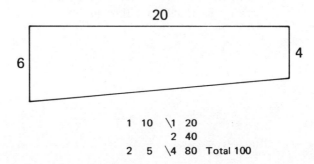

1	10	\1	20	
		2	40	
2	5	\4	80	Total 100

2. Apparently the Egyptians could calculate the areas of fields that were in the shapes of right triangles. Show how they could use this ability to measure the areas of more complicated fields. They could, for example, measure the area of rectangular fields by dividing them into two right triangles.

3. Although the Egyptians had an approximation for π that was good enough for their needs, can you suggest a method they could have used to find a better approximation? Use what you have learned from the preceding question, and remember that the essential properties of π are that for a circle, $C = \pi d$ and $A = \pi r^2$.

***4.** Use what you have learned to investigate the following cuneiform inscription.

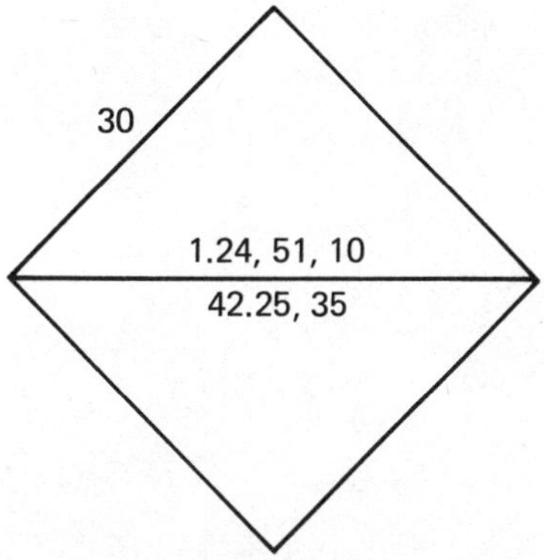

*Discussion questions which are marked with an asterisk require mathematical knowledge not assumed of the average reader.

*5. The Babylonian tables of reciprocals did not contain values for all values of $1/n$. For example, no values were given for $\frac{1}{7}$, $\frac{1}{11}$, $\frac{1}{13}$, and $\frac{1}{14}$. Can you explain why? Can you predict which other numbers would not be included? (*Hint*. Make a table showing the *decimal* values of the reciprocals of the numbers 1 through 20.)

two

the mathematics of the Greeks

In the centuries immediately preceding the birth of Christ, intellectual activity diminished in Egypt and Mesopotamia, and Greece became the center of the civilized world. The Greeks adopted the most advanced scientific methods from both earlier civilizations. The first great Greek mathematicians were those who traveled widely in the neighboring countries. Thales and Pythagoras are the two most famous among them.

Beginnings of the Deductive Method

Thales (*ca.* 600 B.C.) lived in the Greek settlement at Miletus on the eastern shore of the Mediterranean Sea. He is reputed to have been a very wise and clever man; in fact, some say he was the first philosopher. There are many legends attesting that he was a great statesman and businessman as well as a scientist. (There is also a story that says that he once fell down a well while stargazing.)

Thales is thought to have learned mathematics from the Egyptians and probably knew the Babylonian accomplishments in astronomy. He is traditionally regarded as the first true mathematician because of his interest in demonstrating the reasons for geometric truths. In Thales we have the first evidence of a new attitude toward mathematics. After Thales, the Greeks gradually came to regard mathematics as a subject to be studied systematically, logically, and for its own sake rather than to solve a particular real problem.

Some of the theorems attributed to Thales are

THEOREM | The base angles of an isosceles triangle are equal.

THEOREM | When two straight lines intersect, the vertical angles are equal.

THEOREM | If two triangles have two angles and the included side equal, then the remaining sides and angle are equal.

THEOREM | An angle inscribed in a semicircle is a right angle.

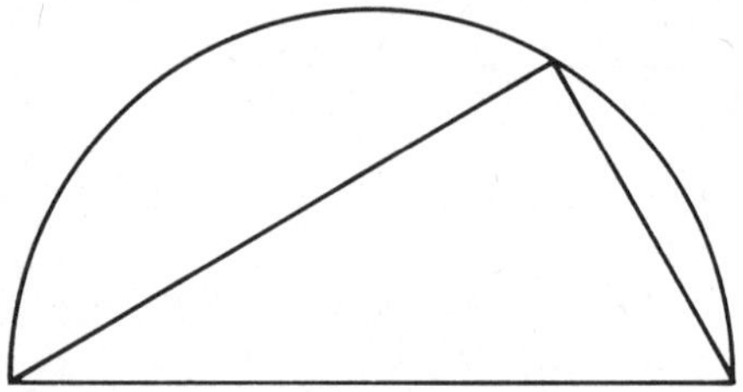

Pythagoras (*ca.* 550 B.C.) was born on the island of Samos not far from Miletus. As a young man he spent many years as a visitor in Egypt and later as a guest (prisoner?) of the Persians in Babylon. During this time he learned both the science and the mysticism of these nations. Upon his return, Pythagoras settled at Crotona in Italy and founded a school. There he taught all subjects, but geometry and the study of numbers were the essential parts of the curriculum.

It is difficult to separate the facts from the legends about the specific mathematical accomplishments of Pythagoras and his students. The Pythagoreans thought that the study of numbers was the only path to a knowledge of all things in the universe. They believed that geometry, although important, was more mundane. They provided proofs for many theorems in both arithmetic and geometry and were almost certainly the first to prove the following two results.

THEOREM | In every triangle the sum of the angles is two right angles.

THEOREM | There is no (rational) number with square equal to 2.

On the other hand there seems to be no direct evidence to indicate that the Pythagoreans proved the theorem that bears their name.

PYTHAGOREAN THEOREM

In a right triangle the square of the hypotenuse equals the sum of the squares of the other two sides.

The Babylonians knew this result many centuries before the Pythagoreans but certainly did not prove it. The Pythagorean theorem was probably first proved by a Greek mathematician about the time that Pythagoras lived. In this sense at least the name of the theorem is appropriate.

The fifth century before Christ was the Golden Age of Greece. It was a time of magnificent accomplishments in history, drama, sculpture, and architecture. The following century saw major advances in philosophy and mathematics. The finest mathematicians in the world were gathered in Athens where Plato had established his Academy. Plato emphasized the importance of mathematics in the training of the mind and had inscribed over the door of his school the words, "Let no one ignorant of geometry enter here." He believed that geometry is not concerned with the physical world but is the study of points, lines, and other quantities as objects of pure reason. Furthermore, Plato stressed the importance of logical order, specific hypotheses, and precise definitions in the study of mathematics. Although he was not in any sense a mathematician, Plato's ideas had a profound influence on the nature of mathematics.

Euclid and the **Elements**

About 300 B.C. Alexander the Great died and his empire was divided. Alexandria in Egypt became the center of learning and it was there that Euclid lived and taught. Almost nothing is known about the life and background of Euclid. He is famous primarily because he compiled all the mathematics known in his time and organized it into the thirteen Books that are collectively called the *Elements*. The first Books are about plane geometry and together constitute the most copied textbook of all time. For more than twenty centuries plane geometry was taught from books that were simply extracts from the *Elements*. It is no wonder that the name of Euclid is so closely associated with geometry.

It should be noted, however, that most of the theorems in the *Elements* are not the original work of Euclid and that the *Elements* contains a large amount of mathematics other than plane geometry. For example, Books VII, VIII, and IX deal with the theory of natural numbers and Book X has an excellent discussion of incommensurable quantities. We shall discuss specific results from several of the Books in later chapters.

Although the *Elements* is valuable as a summary of basic mathematical knowledge, it is the internal structure of the work that makes it so important and the reason that it was so widely imitated. The Greeks developed the axiomatic or deductive form of reasoning and Euclid made full use of this method in the *Elements.* In order to prove a statement by deduction one must show that it is a necessary consequence of a statement that has been verified previously. If one is to avoid circularity and provide a starting point, then some facts must be *assumed.* These statements, from which all others are to be deduced, are called *axioms* or *postulates.* It is also necessary to provide precise definitions for the terms that are used. (This will require, in turn, some undefined terms but Euclid did not formally do this.)

Book I of the Elements begins with a list of twenty-three definitions.[1] The definitions are followed immediately by the assumptions that are to be used. Euclid stated ten axioms, which he divided into two categories.

Postulates

1. A straight line can be drawn between two points.

2. It is possible to extend a finite straight line indefinitely in that straight line.

3. A circle can be drawn with any desired point as center and any finite radius.

4. All right angles are equal.

5. If a straight line intersects two straight lines so that the interior angles on the same side of it sum to less than two right angles, then the two straight lines will intersect if extended indefinitely.

Common Notions

1. Things that are equal to the same thing are also equal to one another.

2. If equals are added to equals, the sums are equal.

3. If equals are subtracted from equals, the remainders are equal.

4. Things that coincide with one another are equal to one another.

5. The whole is greater than the part.

There are many logical defects in Euclid's proofs of the 465 propositions of the *Elements* from these assumptions and many of these defects are major, but the fact remains that the work is essentially correct. We shall consider later some of the vast amount of mathematics that has been developed to understand and correct the errors in Euclid's work.

[1] A list of the definitions is included in the appendix on p. 261.

Archimedes of Syracuse. *The greatest of the ancient mathematicians, Archimedes lived in the Greek city on Sicily. Although he was mainly interested in the general principles of mathematics and mechanics, he was famous during his lifetime for the invention of practical machines.* (*George Arents Research Library*)

Before discussing briefly some of the theorems found in the *Elements,* let us agree on two definitions so that we can save words and effort later.

DEFINITION | Two polygons are **congruent** if corresponding angles are equal and corresponding sides are equal.

DEFINITION | Two polygons are **similar** if corresponding angles are equal and corresponding sides are proportional.

The majority of the theorems in plane geometry in the *Elements* describe properties of straight lines, triangles, parallelograms, and circles. Many propositions state conditions under which polygons will be similar or congruent. Others state consequences deducible from the fact that polygons are similar. The idea of size is central to the entire discussion. Polygons are compared only if they have equal or proportional sides or angles. Of course, there are also many theorems about circles, their chords, tangents, and inscribed angles. As we shall see later, Euclidean geometry can be characterized as the study of the congruence and similarity of plane figures.

Construction Theorems

Euclid made a careful distinction between defining a type of geometric figure and assuming that such objects actually exist. For example, he gives a definition of the special quadrilateral called a square but he does not mention squares again until he proves, at the end of Book I, that such figures can indeed be drawn.

THEOREM (I-46) | A square can be constructed on a given straight line.

The first two of Euclid's axioms ensure that straight lines that are as long as one pleases will exist in geometry. The third axiom states that circles will also exist. Euclid does not *assume* that geometric objects of any other kind can be found in geometry. In particular, the fifth or "parallel" postulate does not imply the existence of parallel lines in Euclidean geometry. Euclid proved that such lines do exist without using the fifth postulate.

THEOREM (I-31) | Through a given point not on a given straight line, a straight line parallel to the given line can be constructed.

The parallel postulate is used to prove the standard facts about parallel lines, including the fact that there is *only one* such parallel line through the given point.

Because of the need to prove the existence of geometric objects, many of the theorems in the *Elements* are of a type called constructions. Such a theorem states that something exists (can be constructed) and the theorem is proved by demonstrating how to draw it. The theorems about squares and parallel lines mentioned above are construction theorems and some other examples from Book I are listed below.

THEOREM (I-1) | On a given finite straight line an equilateral triangle can be constructed.

THEOREM (I-10) | A given finite straight line can be bisected.

THEOREM (I-11) | A perpendicular can be constructed to a given line from any given point on the line.

The propositions of this type have had great appeal to many mathematicians and problem solvers. They consider it a personal challenge to show that "bigger and better" figures can be constructed. A great deal of effort has been expended and some mathematics developed in attempts, some successful and others not, to show that certain constructions can or cannot be done.

Three very famous problems of mathematics, those of trisecting an angle, constructing a square with area equal to that of a given circle, and constructing a cube with volume twice that of a given cube, are of this kind. It would lead us too far afield to consider these problems in detail. One point should be stressed here, however, a point that has not been fully understood by many that have attempted to solve these problems. The problem is not whether these constructions can be done but whether they can be done assuming only Euclid's list of postulates. If one is to abide by these restrictions, then the only tools—either figuratively or literally—that can be used in the constructions are a straightedge for drawing and extending lines and a device for drawing circles. The "straightedge and compass" construction problems are therefore of little practical significance; they have been worked on extensively because of the intellectual stimulation and enjoyment the effort provides.

Summary

During the sixth century B.C. the Greeks began the systematic study of mathematics. They sought the reasons for mathematical truths and attempted to use deduction to prove them. Euclid compiled and organized all the elementary mathematics known in his time in the *Elements*. The origins of many of the important developments and changes in the content and methods of mathematics can be found in the works of the Greeks. In particular, Euclid presented mathematics as a logical system in which the facts are arranged as a sequence of theorems proved from a specific list of assumptions.

EXERCISES

1. Determine which of the theorems of Book I of the *Elements* listed in the appendix state facts about congruence. Simplify the statements of these theorems by rewriting them using the word *congruent*.

2. Verify each of the following statements by finding the Euclidean theorem from which it can be proved directly.

(a) A triangle that has two equal angles is isosceles.
(b) A triangle that has three equal angles is equilateral.
(c) The angles of an equilateral triangle are all equal.

3. Prove Thales' theorem that an angle inscribed in a semicircle is a right angle. (*Hint*. Draw an appropriate radius and then use I-5 and I-32.)

4. As noted above, it is thought that I-32 was first proved by the Pythagoreans. Show how it can be proved using I-29 and I-31.

DISCUSSION QUESTIONS

1. Use Theorem I-32 to discover and prove (informally) a formula for the sum of the angles in a square, pentagon, hexagon, etc.

2. Try to show how the construction theorems in Book I of the *Elements* might be proved by finding a method for drawing the figure using only a straight-edge and compass. Do those theorems mentioned in this section first and then look for others in the appendix. Be careful about making unjustified assumptions about the properties possessed by your figures. Be especially critical and seek justification for your claims about your "square."

3. Is the fifth postulate necessary to prove I-32?

***4.** Show how Euclid probably proved I-1. What assumptions are made that are not included in the axioms?

ree

axiom systems

As implied in the preceding chapter, mathematics today is organized as a structure in which all results are proved from an explicitly stated set of axioms. In order to understand the significance of different branches of mathematics, their interrelationships, and why they were developed, it is necessary to know something about the properties of axiom sets. In this section we shall define the properties of *consistency, equivalence,* and *independence* for axiom sets and illustrate them with some examples of axiom systems. In order that the discussion remain as concrete as possible, these ideas will be introduced in terms of their applicability to the set of axioms Euclid used in the *Elements.* A detailed investigation of the properties of Euclid's axioms constitutes a major part of the following chapters on geometries.

Following the tradition begun by the Greeks, when mathematics is done formally and with attention to the requirements of logical rigor, it is organized according to the outline shown below. (We shall not often be rigorous in our look at mathematics. Instead we shall be content with informal discussions of the formal structure.)

1. A set of words which are chosen to be **undefined terms** and from which all other terms are defined as needed.

2. A set of statements which are chosen as unproved assumptions and from which all other statements are to be proved. These statements are called **axioms** or **postulates.**

3. A set of statements which can be proved from the axioms. These statements are called **theorems.**

We shall call any set of things that conforms to this pattern an **axiom syste**

Euclid tried to define all the mathematical terms he used in t *Elements*. If we suppose that the words *point, line, on* in the sense that point is *on* a line, and a few others are undefined, then the *Elements* contai the development of a large quantity of mathematics as an axiom syste The geometry which is obtained from the axioms of the *Elements* is call Euclidean geometry to distinguish it from other geometries which are bas on different sets of axioms.

Consistency

Aside from the value it may have as an exercise in logic, the stu of an axiom system becomes worthwhile only when it contains theore that are of interest when some meanings are assigned to the undefined tern Euclid believed that each of his axioms was recognizable as an obvious tru about the real world. He thought that in his "definitions" he had assign to his terms meanings which gave the essential and idealized properties objects which actually exist.

It is possible that an axiom system contain two theorems th contradict each other. If it were possible to assign meaning to the undefin terms of such a system in a way that made all the axioms true, then bo theorems would necessarily be true and a contradiction would exist in natu As far as is known, however, no contradictions exist in the real worl Therefore, a system that contains contradictory theorems has no applicatio

DEFINITION | A set of axioms is said to be **consistent** if it is not possible to prove contr dictory theorems from its axioms.

DEFINITION | A **model** for an axiom system is the result of assigning specific meanin to the undefined terms in such a way that all the axioms become tr statements.

Because of our belief that no contradictions exist in the real wor it is possible to show that a set of axioms is consistent by providing a mo for the system. Euclid was certain that his system had the geometry of t real world as its model and was therefore obviously consistent. The ea mathematicians agreed with Euclid's view and it was many centuries bef anyone thought it necessary to provide a more careful verification of t consistency of Euclid's axioms.

Equivalence

Mathematicians did raise one objection to Euclid's set of axioms almost immediately. They felt that his fifth postulate, the one about parallel lines, was unnecessarily complicated and far from obvious. It was deemed desirable to replace this axiom with some other statement which would be easier to understand but which would result in a system with exactly the same theorems.

DEFINITION | Two axiom systems with the same undefined terms are said to be **equivalent** if all the axioms of each system can be proved as theorems in the other system.

Thus, what was being sought was a replacement for the fifth postulate that would result in an equivalent system. The search for such substitutes was successful. The axiom that has been most widely used in place of Euclid's parallel postulate bears the name of one of the relatively modern mathematicians who advocated its use, John Playfair (1748–1819). It was actually first suggested and used by the Greeks soon after Euclid's death.

Playfair Axiom: Through a given point not lying on a given line there is exactly one line parallel to the given line.

Further examples of statements that can be used in place of the parallel postulate will be discussed in the next chapter. (At this point it would be worthwhile to compare the Playfair Axiom with the comments made about I-31 and the fifth postulate on page 16.)

Dependence and Independence

A number of mathematicians sought to solve the difficulties with the fifth postulate in an entirely different way. They suggested that perhaps the fifth postulate was redundant and that an equivalent system would result if that statement were simply not included in the axioms. That is, they conjectured that all the proofs of theorems that seem to require the use of the fifth postulate could actually be done using only the other axioms in Euclid's system. If their conjecture were true, then it would be possible to prove the fifth postulate as a theorem in the system obtained by omitting it as an axiom.

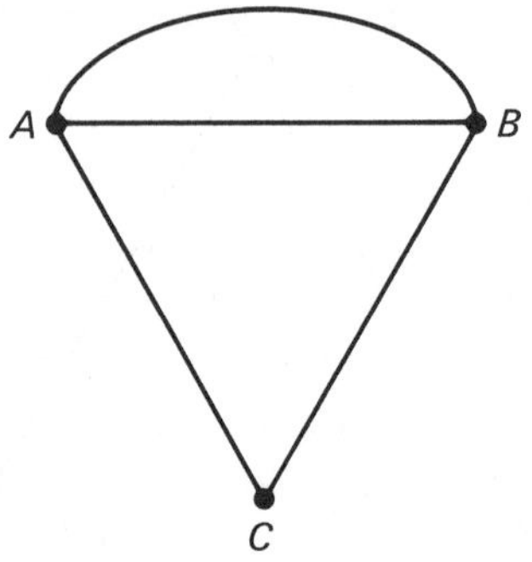

Axiom 3

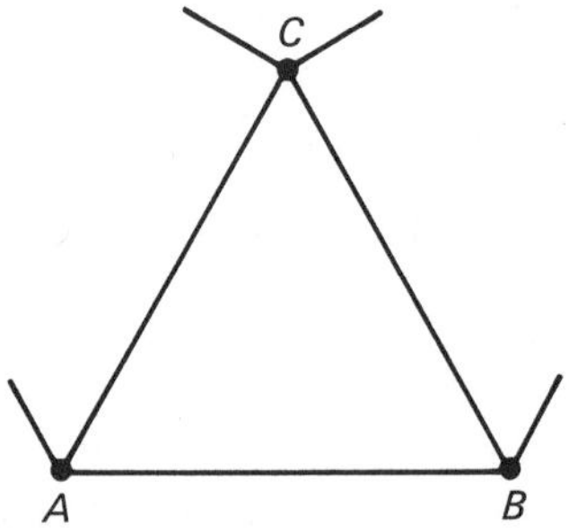

Axiom 4

Example II: A more interesting mathematical system results from usi the following set of axioms.

A1. There exists at least one line.

A2. There are exactly three points on every line.

A3. For every line there is at least one point not on the line.

A4. Any two points lie on exactly one line.

A5. There is at least one point on any two distinct lines.

Attempts to find a model for this axiom system lead us immediate to the conjecture and proof of some theorems.

THEOREM 1 | There is exactly one point on any two distinct lines.

Proof: By A5 there is at least one point on two distinct lines. If there we two (or more) distinct points on two distinct lines, it would contradict A

THEOREM 2 | There are at least seven points.

Proof: Let l_1 be a line with points A, B, and C, which we shall denc $l_1 = [A, B, C]$, and let D be a point not on l_1. All of this is possible by t first three axioms.

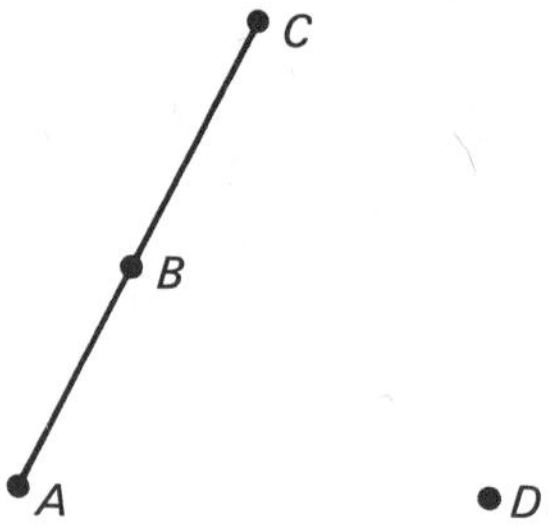

By A4 there must be a line containing B and D. Then there must be a third point E on this line by A2. Since D is not on l_1, this new line $l_2 = [B, D, E]$ is distinct from l_1. By Theorem 1 we see that E is distinct from A and C.

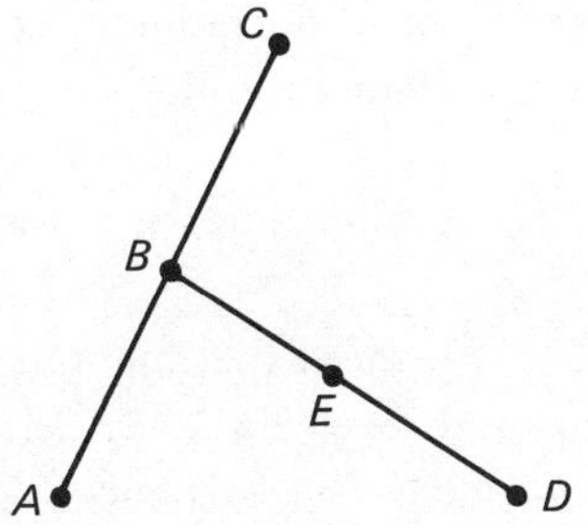

Similarly, the points C and D determine a third line $l_3 = [C, D, F]$ that is not l_1 or l_2. Applying Theorem 1 again we see that the point F is distinct from the others. Using the argument one more time with the points C and E we obtain a seventh point G.

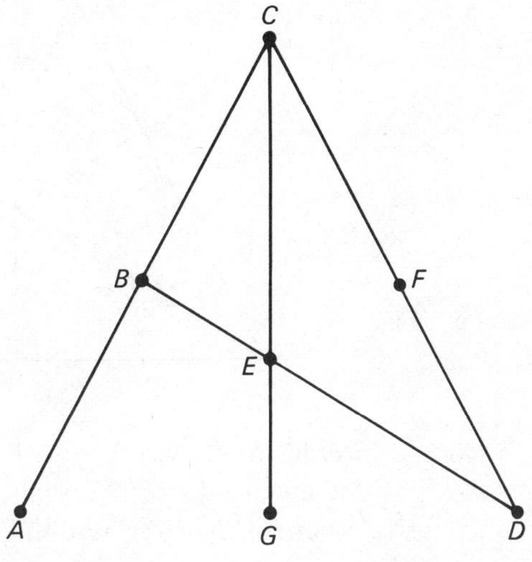

THEOREM 3 | There are exactly seven points.

Proof: From Theorem 2 we have found seven points A, B, C, D, E, F, G and four lines, $l_1 = [A, B, C]$, $l_2 = [B, D, E]$, $l_3 = [C, D, F]$, $l_4 = [C, E, G]$. Suppose there is another point H. Then A and H determine a line which must contain a third point I.

Consider the line $[A, H, I]$. By comparing this new line with l_1 an
using Theorem 1, we see that I is different from both B and C.

The line $[A, H, I]$ must have a point in common with each of th
lines l_2, l_3, and l_4. The common point must be I and from consideratio
of $l_2 = [B, D, E]$ we see that I is D *or* I is E. But if I is E, then $[A, H$
$I]$ and l_3 have no point in common and if I is D, then $[A, H, I]$ and l
have no common point.

The assumption that an eighth point H exists leads to contradictions
The assumption must be false and the theorem true.

Further experimentation will eventually provide a model with seven
points and seven lines (BFG is a line). It is easily checked that the axiom
are satisfied. The existence of the model shows that the set of axioms i
consistent.

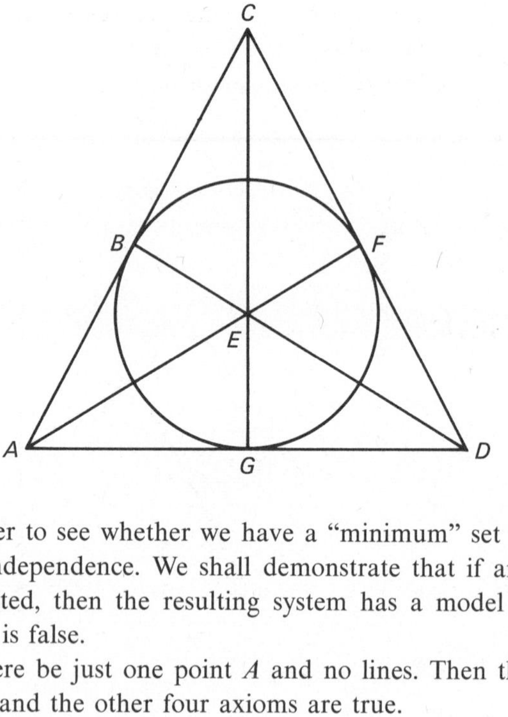

In order to see whether we have a "minimum" set of axioms, le
us check for independence. We shall demonstrate that if any one of the
axioms is omitted, then the resulting system has a model in which the
omitted axiom is false.

Let there be just one point A and no lines. Then the first axiom
is clearly false and the other four axioms are true.

A2 is independent since the model for Example I satisfies A1, A3
A4, and A5 and does not satisfy A2.

If we take $l_1 = [A, B, C]$ as our only line and points, then A3 is
false but the other axioms are true.

The models represented by the figures below show that A4 and A5, respectively, are not dependent on the others.

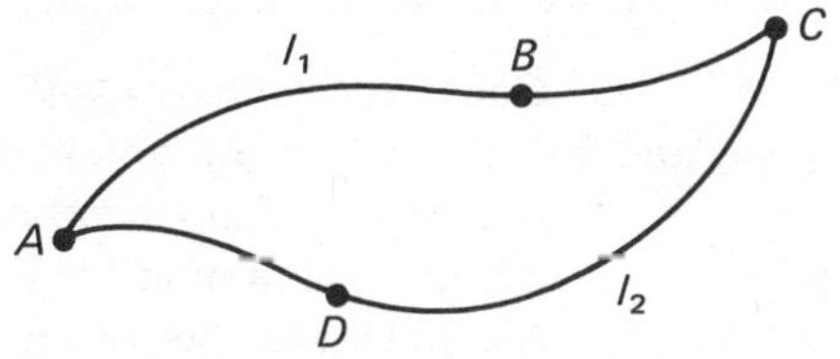

A	A	A	A	B	B	B	C	C	C	D	G
B	D	E	F	D	E	F	D	E	F	E	H
C	G	I	H	I	H	G	H	G	I	F	I

Summary

The set of axioms for an axiom system must be consistent if the system is to have any applications in the real world. It is of interest to know whether the axioms are independent and to know what other sets of axioms will provide an equivalent axiom system. The attempts to provide Euclidean geometry with a consistent and independent set of axioms led to the development and study of other axiom systems.

EXERCISES

1. Take as undefined terms the words *point, line,* and *on*. Use the following four statements as the set of axioms. Is this system consistent? Are the axioms independent? Answer the same questions for each of the four systems obtained by omitting one axiom.

A1. There are exactly three points.
A2. All points lie on the same line.
A3. Any two points lie on exactly one line.
A4. There are at least two lines.

2. (a) Is it true in general that if an axiom is omitted from a consistent set of axioms, then the resulting set of axioms is also consistent?

(b) Is the preceding statement true if the word *consistent* is replaced by the word *independent*?

(c) If the same axiom is added to each of two equivalent systems, are the resulting systems equivalent?

(d) If two axiom systems are equivalent and one of them i consistent is the other system also consistent?

(e) Is the preceding statement true when the word *consistent* i replaced by the word *independent*?

3. Consider the system with undefined terms *club, committee, mem ber,* and *belongs to* and the following three axioms.

A1. Exactly four members belong to the club.

A2. Each committee consists of exactly two members of the club

A3. Each member of the club is on exactly one committee wit each other member of the club.

(a) Show that the axioms are consistent.

(b) Show that the axioms are independent.

(c) Show that there are exactly six committees in this system

(d) Show that each member not on a given committee belong to exactly one committee that has no member in common with th given committee.

***4.** Replace the fifth axiom of the seven point geometry with "Give a line and a point not on the line, there is exactly one line through th point that does not intersect the given line." Check for independence an consistency. Prove as many theorems as you can. (*Hint*. See the mode used for the independence of the original *A*5.)

DISCUSSION QUESTIONS

1. Explain why adding more axioms to an axiom system will, i general, increase the number of theorems and decrease the number o models.

2. If the "real world" is a model for Euclidean geometry, then al Euclid's axioms and theorems should be true and applicable to practica problems. Do you think that this is the case? Notice particularly whethe Theorem I-32 is true for triangles on the surface of Earth.

part two

geometry

ɔur

ɑbsolute geometry ɑnd the parallel postulate

In this section we begin our investigations of the developments in geometry since the time of the *Elements*. The first topic of the discussion is the consequences of the attempts to deal with the alleged deficiences in Euclid's fifth postulate about parallel lines.

As mentioned in the preceding chapter, many early mathematicians thought this axiom to be at best more complicated than necessary and at worst dependent on the other nine axioms. The desire for a "more obviously true" axiom led to a search for equivalent statements. The conjecture of dependence caused numerous attempts at providing a proof for the fifth postulate.

These two ventures are closely related. The parallel postulate could be proved dependent by proving any one of the equivalent statements dependent. Many statements equivalent to the fifth postulate were found but no one was ever able to prove any of them dependent on Euclid's other axioms. The attempts did lead to the discovery of many important mathematical results, however, and even contributed to the creation of entirely new branches of mathematics.

Absolute Geometry

The entire investigation of the fifth postulate, its equivalents, and their possible dependency, must begin with a study of the properties of the system that has Euclid's other nine assumptions as its set of axioms.

DEFINITION | The geometry obtained by using as axioms all Euclid's axioms except th fifth postulate is called **absolute geometry.**

Clearly any theorem of Euclidean geometry that can be prove without using the parallel postulate is a theorem in absolute geometr Conversely, any theorem in absolute geometry is also a theorem in Euclidea geometry and in any other system obtained by adding new axioms to thos of absolute geometry.

Many of the theorems in Book I of the *Elements* are also true i absolute geometry. The first twenty-eight (I-1 through I-28) are example These theorems deal mainly with the properties of triangles. They show th triangles can be constructed, give facts about the relative size of angles an sides, and specify conditions under which triangles will be congruent.

Theorems I-27 and I-28 are much like each other. Both state cond tions under which lines will be parallel and both are theorems in absolut geometry.

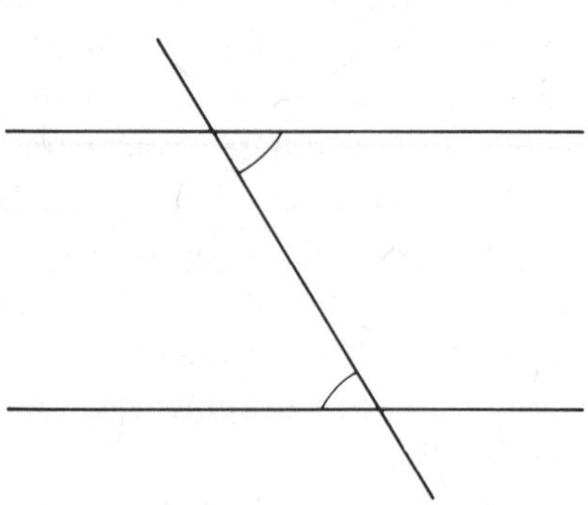

THEOREM (I-27) | If a straight line intersects two straight lines so that the alternate angle are equal, then the two straight lines are parallel.

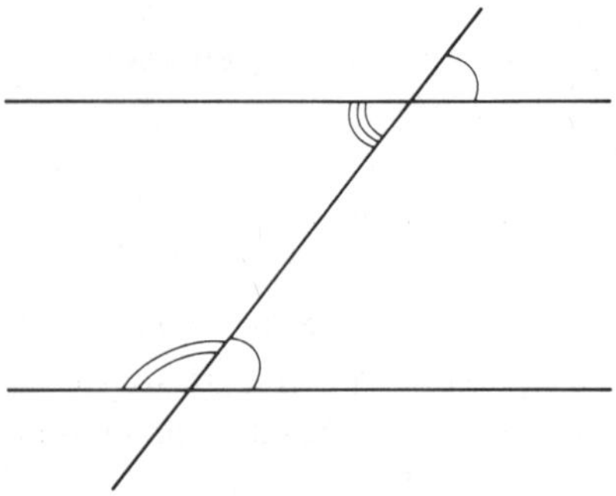

THEOREM (I-28) If a straight line intersects two straight lines so that the exterior angle is equal to the opposite interior angle on the same side or if the two interior angles on the same side are equal to two right angles, then the two lines are parallel.

Notice that these theorems do not say that parallel lines exist. From either of them, however, it is possible to prove that parallel lines exist in absolute geometry.

THEOREM (I-31) Through a given point not on a given line a line parallel to the given line can be constructed.

The fifth postulate is first used by Euclid to prove Theorems I-29, I-30, and I-32. The last two of these are of particular interest.

THEOREM (I-30) Straight lines parallel to the same straight line are parallel to each other.

THEOREM (I-32) The sum of the angles in any triangle is equal to two right angles.

The statement of each of these two theorems is equivalent to the Playfair Axiom and therefore equivalent to Euclid's parallel postulate. More precisely, the four systems that result when each of these statements is added to the set of axioms for absolute geometry are all equivalent. If any one of the four statements could be proved as a theorem in absolute geometry, then all four could be proved and each would be dependent on the axioms of absolute geometry. In such an event it would have been proved that absolute geometry and Euclidean geometry are equivalent.

Saccheri's "Endeavor"

In the eighteenth century several mathematicians made attempts to prove the fifth postulate that were to directly influence the course of mathematical progress since that time. One man who believed that he had found a proof was Girolamo Saccheri (1667–1733). Saccheri was a Jesuit and professor of mathematics at the University of Pavia. In the year of his death he published his work in a book that was titled (approximately) "Euclid Freed of Every Defect or A Geometric Endeavor in Which are Established the Foundation Principles of Universal Geometry." In the preface Saccheri explained that while no one doubted the excellence and

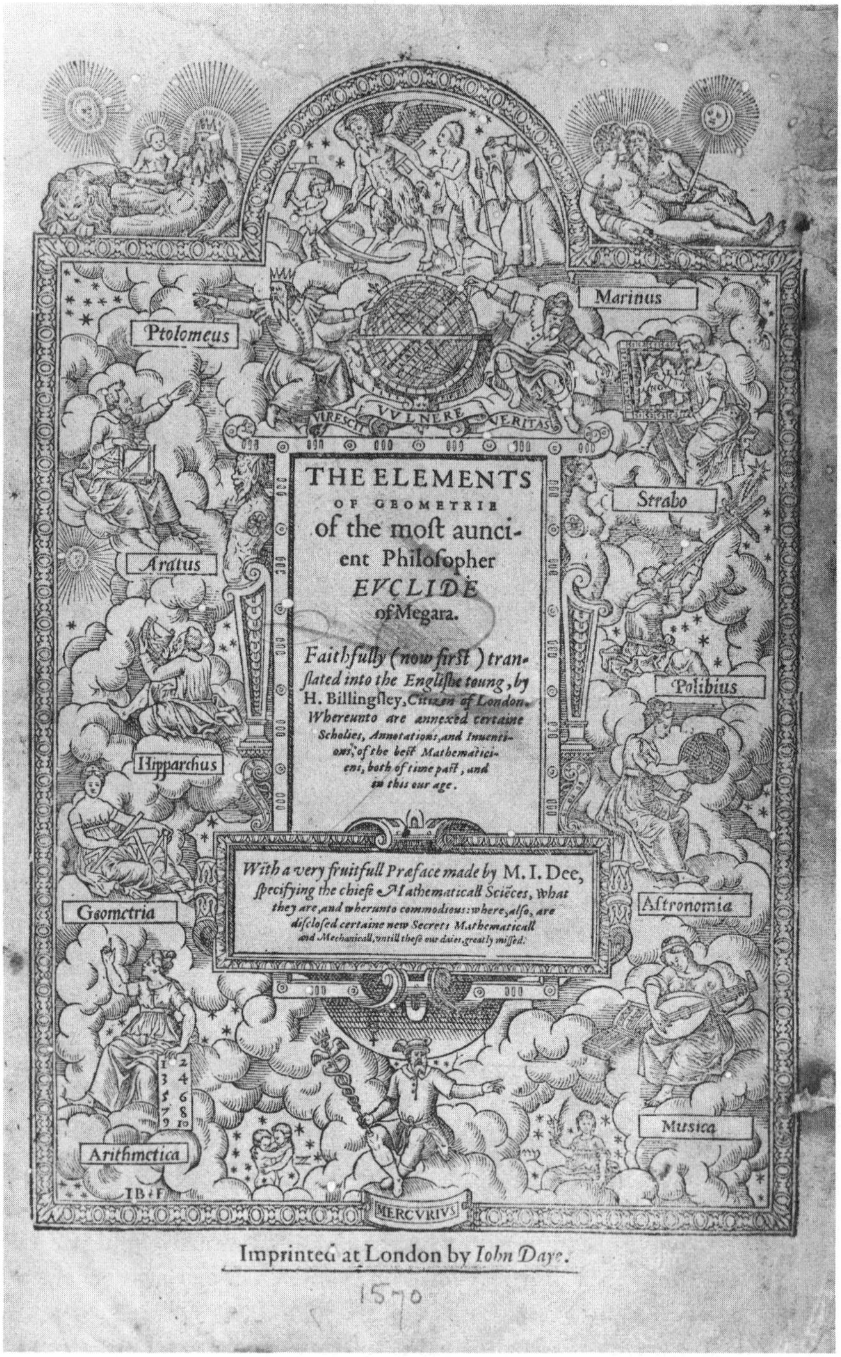
THE ELEMENTS
OF GEOMETRIE
of the most aunci-
ent Philosopher
EVCLIDE
of Megara.
Faithfully (now first) tran-
slated into the Englishe toung, by
H. Billingsley, Citizen of London.
Whereunto are annexed certaine
Scholies, Annotations, and Inventi-
ons, of the best Mathematici-
ens, both of time past, and
in this our age.
With a very fruitfull Praface made by M. I. Dee,
specifying the chiefe Mathematicall Sciēces, what
they are, and wherunto commodious: where, also, are
disclosed certaine new Secrets Mathematicall
and Mechanicall, untill these our daies, greatly missed.
Ptolomeus
Marinus
Aratus
Strabo
Hipparchus
Polibius
Geometria
Astronomia
Arithmetica
Musica
VIRESCIT VULNERE VERITAS
MERCVRIVS
Imprinted at London by Iohn Daye.
1570

***Euclid's* Elements.** *Title page of the first English publication of Euclid's* Elements. *The translation is believed to have actually been done by John Dee.* (*George Arents Research Library*)

worth of Euclid's work, there were a few parts of his logical exposition that needed to be improved.

In particular it was Saccheri's goal to simplify Euclid's set of axioms by proving the parallel postulate dependent upon the others. He correctly proved a large number of theorems and the error in his work occurs only at the very end of the book. Saccheri died believing that he had accomplished his purpose of correcting minor flaws in Euclid. What he had actually done is lay the groundwork for an entirely new geometry in which the theorems contradict those of Euclidean geometry.

In order to make it easier to explain Saccheri's theorems and their proofs, let us agree on the following notation. Capital letters A, B, C, . . . will be points. Pairs of points, AB, AC, . . . , will represent the line segment between the points, and $|AB|$ will be the length of the segment AB. The angle of less than two right angles with vertex at B and having A and C on the rays will be represented $\angle ABC$ or $\angle CBA$ or simply $\angle B$ if there is no risk of confusion. The size of $\angle ABC$ will be denoted by $|\angle ABC|$. When we refer to Euclid's third postulate or the eighth theorem of Book I of the *Elements,* we shall write simply P3 or I-8. Our proofs of theorems will be informal since logical rigor is neither desirable nor practical at this point.

Saccheri chose to prove the fifth postulate dependent by proving an equivalent statement. To this end he defined a special kind of quadrilateral.

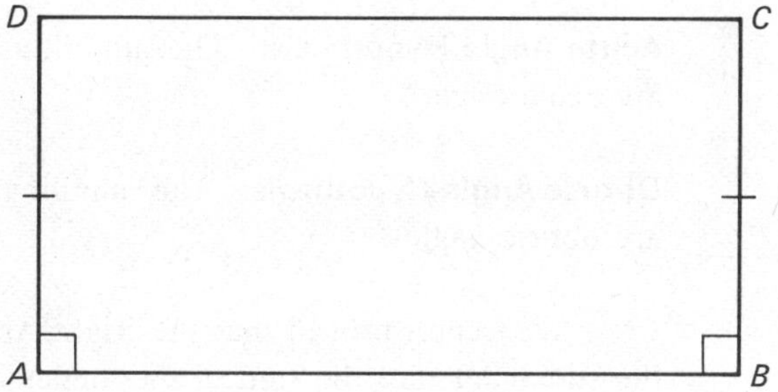

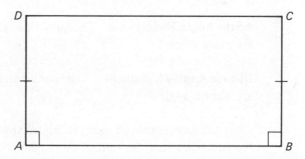

DEFINITION | A quadrilateral $ABCD$ is called a **Saccheri quadrilateral** if angle A and angle B are right angles and $|AD| = |BC|$.

Saccheri then proved the following theorem in absolute geometry.

Proposition I: If $ABCD$ is a Saccheri quadrilateral, then angles C and D are equal.

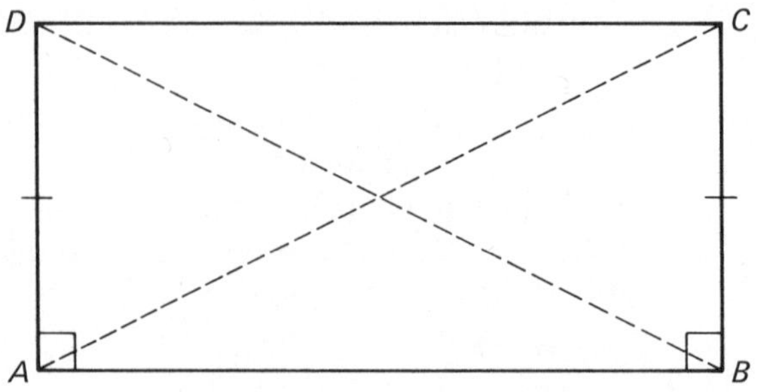

Proof: Draw AC and BD (P1). The triangles ABC and ABD are congruent (I-4). In particular $|\angle ACB| = |\angle ADB|$ and $|AC| = |BD|$. The triangles ACD and BCD are congruent (I-8). In particular, $|\angle ACD| = |\angle BDC|$. Thus, $|\angle C| = |\angle D|$ (Common Notion 2).

It follows from Proposition I that there are just three possible cases that can occur. The "summit" angles C and D are both acute, both obtuse, or both right angles. Saccheri proved that if one of these cases occurs in some quadrilateral of this kind, then that same case occurs in every such quadrilateral.

Right Angle Hypothesis: The summit angles of a Saccheri quadrilateral are right angles.

Acute Angle Hypothesis: The summit angles of a Saccheri quadrilateral are acute angles.

Obtuse Angle Hypothesis: The summit angles of a Saccheri quadrilateral are obtuse angles.

Saccheri proved that the Right Angle Hypothesis is equivalent to the statement that the sum of the angles in a triangle is two right angles and therefore equivalent to Euclid's P5. His goal was to prove the Right Angle Hypothesis as a theorem in absolute geometry. He tried to do this by showing that the two alternative hypotheses could be eliminated. Moré precisely, he attempted to show that the two systems obtained by using the Acute Angle Hypothesis and the Obtuse Angle Hypothesis as an additional axiom with those of absolute geometry are inconsistent. Saccheri reasoned that the Right Angle Hypothesis must be a theorem if assuming that it is false leads to contradictions.

Saccheri found the system that uses the Obtuse Angle Hypothesis to be the easier to work with. By using theorems of absolute geometry and the assumption that the summit angles are obtuse he was able to prove the following theorem.

THEOREM | If a straight line intersects two straight lines in such a way that the internal angles on the same side are less than two right angles, then the two straight lines will intersect at some point on the same side as the angles.

This theorem is exactly Euclid's fifth postulate. It had been shown to be equivalent to the Right Angle Hypothesis and thus contradictory to the Obtuse Angle Hypothesis. Saccheri had proved that the system obtained by using the Obtuse Angle Hypothesis is inconsistent.

The Acute Angle Hypothesis

Saccheri's attempts to show that the system obtained by using the Acute Angle Hypothesis as an axiom were less successful. It will be instructive to consider briefly the method he used because it is analogous to the one by which he achieved a contradiction to the Obtuse Angle Hypothesis. Saccheri had proved the following theorem of absolute geometry.

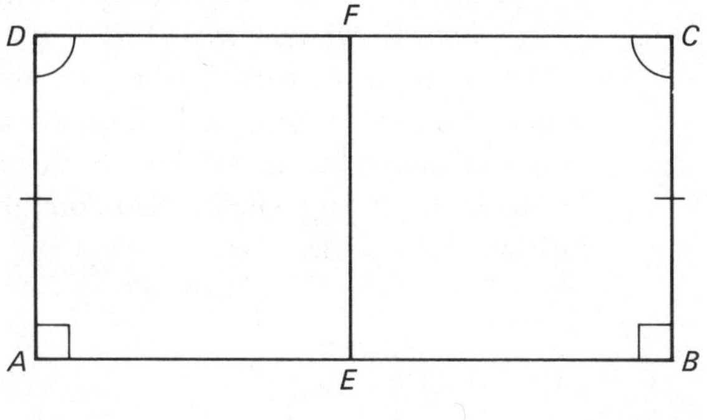

Proposition II: If $ABCD$ is a Saccheri quadrilateral with E the midpoint of AB and F the midpoint of CD, then the angles AEF, BEF, CFE, and DFE are right angles.

Proof: Draw EF, AF, BF, CE, and DE. Then triangle ADE is congruent to triangle BCE (I-4). In particular $|CE| = |DE|$. Thus triangles CEF and DEF are congruent and $|\angle DFE| = |\angle CFE|$ (I-8). Also, triangle ADF is congruent to triangle BCF (Proposition I and I-4) so that $|AF| = |BF|$. Thus triangles AEF and BEF are congruent and $|\angle AEF| = |\angle BEF|$. The required angles are therefore all right angles by definition.

Saccheri made use of the Acute Angle Hypothesis and began to try to find theorems that would show the system to be inconsistent.

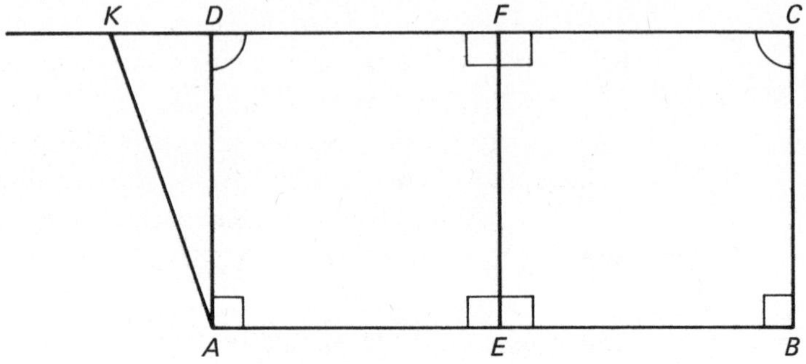

Proposition III: If $ABCD$ is a Saccheri quadrilateral, then, $|CD| > |AB|$.

Proof: Bisect AB and CD and let these points be E and F (I-10). Draw EF. If $|DF|$ were equal to $|AE|$, then $EFDA$ would be a Saccheri quadrilateral (Proposition I) and by the Acute Angle Hypothesis angle A would have to be acute. Since $\angle A$ is a right angle, it must be that $|DF| \neq |AE|$.

Extend CD past D (P2). If K is any point on the extension, then $\angle EAK$ is obtuse. In particular, if $|DF|$ were less than $|AE|$, then K could be put on the extension so that $|KF| = |AE|$ (I-3). This would make $EFKA$ a Saccheri quadrilateral. But then the fact that $\angle EAK$ is obtuse contradicts the Acute Angle Hypothesis. Therefore, it must be that $|DF| > |AE|$ and thus that $|CD| > |AB|$.

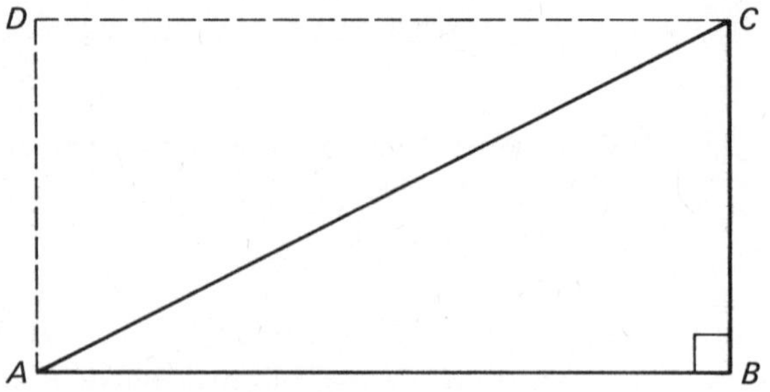

Proposition IV: In any right triangle the sum of the other two angles is less than a right angle.

Let ABC be a right triangle with $\angle B$ a right angle. At A, construct a line perpendicular to AB (I-11). On this perpendicular, construct a segment of length $|BC|$ with one end at A so that the other endpoint, D, is not on the same side of AC extended as B (I-2). Draw DC (P1). Then $ABCD$ is a Saccheri quadrilateral. By Proposition III we have $|CD| > |AB|$. Therefore, $|\angle ACB| < |\angle CAD|$ (I-25). Thus, $|\angle ACB| + |\angle BAC|$ is less than $|\angle CAD| + |\angle BAC|$, which is the size of the right angle A.

These last two theorems contradict theorems of Euclidean geometry. They do not, however, contradict any of the axioms or theorems of the system with which Saccheri was working. After proving a number of results of this kind, Saccheri was overwhelmed by the "absurdity" of his results. He was unable to find a contradiction in the system but he was convinced that there was no way that such strange theorems could possibly describe the real world. He concluded his investigation by giving an argument in support of the following statement.

Proposition XXXIII: The Acute Angle Hypothesis is inconsistent because it is repugnant to the nature of straight lines.

Saccheri's abandonment of logical rigor in favor of intuition was caused by his conviction that geometry must describe reality and that Euclidean geometry does so. Although the use of the Absolute Angle Hypothesis does enable one to prove many statements that do not conform to our intuitive ideas about the real world, it will nevertheless result in a consistent logical system.

Summary

Saccheri tried to prove that Euclid's fifth postulate is dependent upon Euclid's other axioms. He showed that the Right Angle Hypothesis is equivalent to the parallel postulate and attempted to prove that the alternative systems, obtained by adding the Obtuse Angle Hypothesis and Acute Angle Hypothesis as axioms to the axioms of absolute geometry, are inconsistent. Saccheri was able to prove that the Obtuse Angle Hypothesis results in an inconsistent system but he was not able to reach a contradiction to the assumption of the Acute Angle Hypothesis. (He thought he had done it; hence the title of his book.)

Saccheri's work was largely ignored or unknown and other mathematicians later duplicated his efforts. It was not until the nineteenth century that it was realized that a geometry other than Euclid's might be consistent. The new geometry was at that time discovered and developed independently by Carl Friedrich Gauss in Germany, by Janos Bolyai in Hungary, and by Nicolai Lobatchevsky in Russia.

EXERCISES

1. Prove the following propositions in the geometry obtained by adding the Acute Angle Hypothesis to the postulates for absolute geometry. (*Hint*. Use Proposition IV as a starting point.)

I. The sum of the angles in any triangle is less than two right angles.

II. The sum of the angles in any quadrilateral is less than four right angles.

III. An angle inscribed in a semicircle is acute. (*Hint*. Use I-5.)

2. Explain how it could be proved that from the Acute Angle Hypothesis it follows that the sum of the angles in an n-sided polygon is less than $2(n - 2)$ right angles.

3. Prove that in absolute geometry there exists at least one line parallel to a given line through a given point (I-31). [*Hint:* Use I-27 (p. 32) and I-23 (p. 263). Start by drawing a line between the given point and line.]

4. Prove that the statement "Through a given point not on a given line, there is at most one line parallel to the given line." is equivalent to the Playfair Axiom (when added to the axioms of absolute geometry).

5. Prove that the Playfair Axiom and I-30 are equivalent. (*Hint:* Show that if either is true, then the assumption that the other is false leads to a contradiction.)

DISCUSSION QUESTION

Consider again the theorems of Euclid's *Elements*. Determine which of these results are not true if one assumes the Acute Angle Hypothesis instead of the fifth postulate and conjecture how they could be changed to obtain theorems in this system. Make a list of Euclidean results and their analogues in the *acute angle* geometry.

five

non-Euclidean geometries

Carl Friedrich Gauss (1777–1855) was probably the first man to realize that a geometry using axioms other than those of Euclid is possible. Gauss was the son of a farmer and would not ordinarily have received more than a very minimal formal schooling. His mathematical genius was apparent, however, while he was still a young boy and a nobleman became his patron and financed his education. Gauss made several significant contributions to mathematics even before he entered college and his doctoral dissertation in 1799 is a landmark in algebra. (An excellent account of his life is contained in E. T. Bell's *Men of Mathematics*.)

Gauss was a perfectionist and never published his work unless he was certain that he had the complete answer in the most elegant and precise form possible. His motto was "pauca sed matura," which means "few, but ripe." Many of his results, including his work on this geometry, were not published in his lifetime. Gauss also realized that although the first man to show the existence of a new geometry would eventually receive high honors and fame, the public and many mathematicians would be quick to severely criticize that man as a fool. He had no need for additional fame and left the official recognition for the next to discover the result.

Janos Bolyai (1802–1860) was a Hungarian cavalry officer and the son of Wolfgang Bolyai, a friend and fellow student of Gauss. As a young man he was interested in mathematics and worked on the problem of the Euclidean parallel postulate even though his father had warned him that such investigations would almost certainly end in failure. He did of course fail to prove the postulate dependent but he saw the alternative and developed, about 1824, a new geometry equivalent to that unwittingly begun by Saccheri. After many delays his work was finally published as an appendix to a book by his father in 1832. He titled it "The Absolute Science of Space."

The young Bolyai did not live long enough for the world to fully realize what he had created. He died embittered after what little recognition there was for the achievement during his lifetime was privately claimed by and given to a German, Gauss, and the official recognition was awarded to a Russian.

Lobatchevskian Geometry

Nicolai Ivanovitch Lobatchevsky (1793–1856) was a full professor at the University of Kazan by the time he was 23 years old. He taught and made contributions to mathematics, physics, and astronomy. He also ran the library and the museum and supervised the students. Eventually he became head of the University.

Beginning about 1820 Lobatchevsky studied the theory of parallel lines. After a few years he realized that it was fruitless to try to prove the parallel postulate and in 1826 he published a memoir on the subject. This paper contained the development of what Lobatchevsky called "imaginary geometry." This geometry is essentially the same as that discovered by Bolyai and Saccheri. Although it was many years before the value of his work was finally understood, Lobatchevsky's contributions were eventually rewarded by the decision to describe this geometry as Lobatchevskian Geometry.

Lobatchevskian Geometry is the system obtained by adding to the axioms of absolute geometry the following statement.

Lobatchevsky's Axiom: If A is a point not on the line BC and D is the point on BC such that $AD \perp BC$, then there are rays AH and AK from A that are not on the same line, which do not intersect BC, and such that any line AF within the $\angle HAK$ containing D does intersect BC and any line AG that does not pass through $\angle HAK$ does not intersect BC.

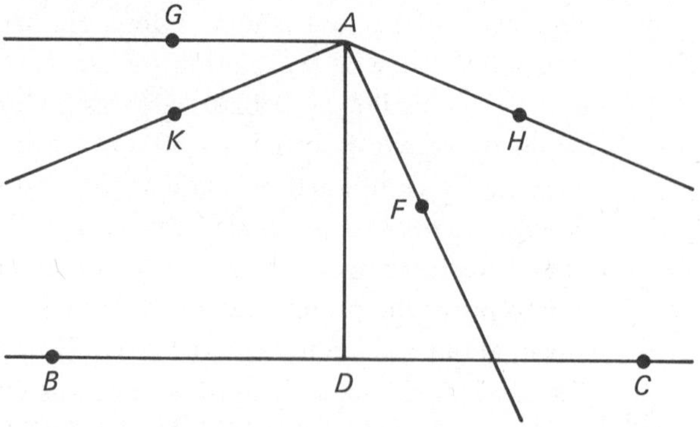

An extensive discussion of the history of the invention of this geometry and translations of the basic works of Bolyai and Lobatchevsky can be found in the book by Roberto Bonola.[1] We shall content ourselves here with a look at a few of Lobatchevsky's first theorems and definitions.

DEFINITION | The lines AH and AK are called **parallels** to BC through A. The line AH is the parallel to the directed line $\overrightarrow{BC}$ and AK is the parallel to the directed line $\overrightarrow{CB}$.

THEOREM L1 | If A is not on the line BC, D is on BC so that $AD \perp BC$, and AH, AK are the parallels to BC through A, then $\angle HAD$ and $\angle KAD$ are equal acute angles.

Proof: Suppose one of the angles is larger than the other, say, $|\angle HAD| > |\angle KAD|$. Construct within $\angle HAD$ the line AF so that $|\angle FAD| = |\angle KAD|$. Then AF intersects BC at some point L. Construct on BC, on the opposite side of D from L, a segment DE so that $|DE| = |DL|$. Draw EA.

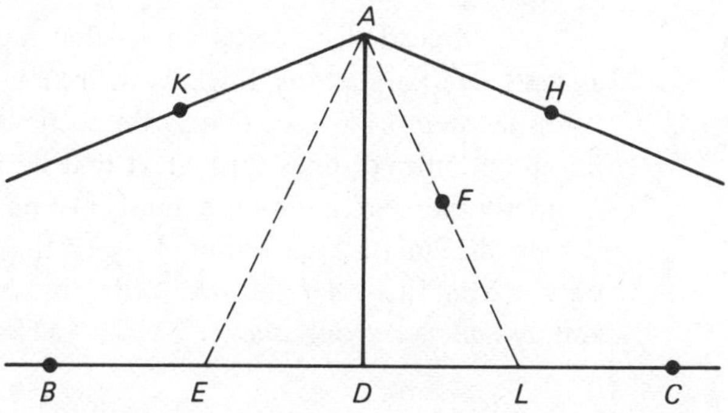

Then triangle EAD is congruent to triangle LAD, and in particular $|\angle LAD| = |\angle EAD|$. But then AK and AE coincide. This is impossible since AK does not intersect BC. Thus, $|\angle HAD| \not> |\angle KAD|$. Similarly, $|\angle KAD| \not> |\angle HAD|$. Thus, $|\angle HAD| = |\angle KAD|$. These angles are not right angles because Lobatchevsky's Axiom implies that AH and AK are not on the same line. These angles are not obtuse because if they were, then the line at right angles to AD through A would intersect BC and this would contradict the absolute theorem I-28.

[1] Roberto Bonola, *Non-Euclidean Geometry: A Critical and Historical Study of its Developments* (New York: Dover Publications, 1955).

DEFINITION | The acute angle that *AH* and *AK* make with *AD* is called the **angle of parallelism** at *A* for the line *BC*.

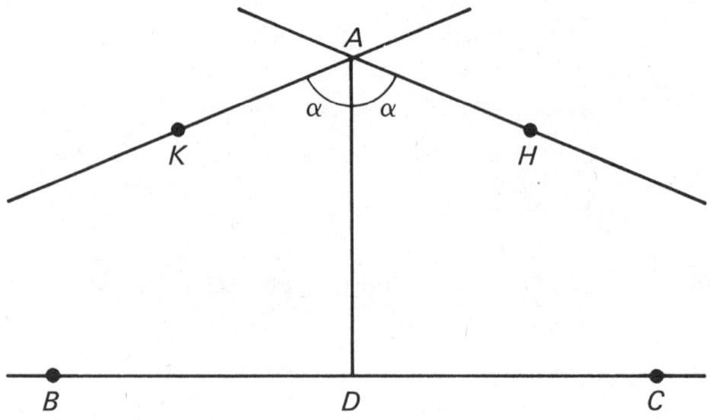

As pointed out before, the first twenty-eight of the propositions of the *Elements* were proved without using the parallel postulate. These theorems of Euclidean geometry are therefore theorems of absolute geometry and also of Lobatchevsky's geometry, which is an extension of absolute geometry.

Proposition I-28 guarantees that in the absolute geometry there is at least one parallel, by Euclid's definition of parallel, to *BC* through *A*. From theorem L1 we see that in Lobatchevsky's geometry there will be an infinite number of lines through *A* that do not intersect *BC*. Thus, in this geometry there are an infinite number of parallels to a line through a point not on the line. In the terminology of this geometry, however, there are only two parallels, *AH* and *AK*, to *BC* through *A*. The other lines are called simply *nonintersecting lines*.

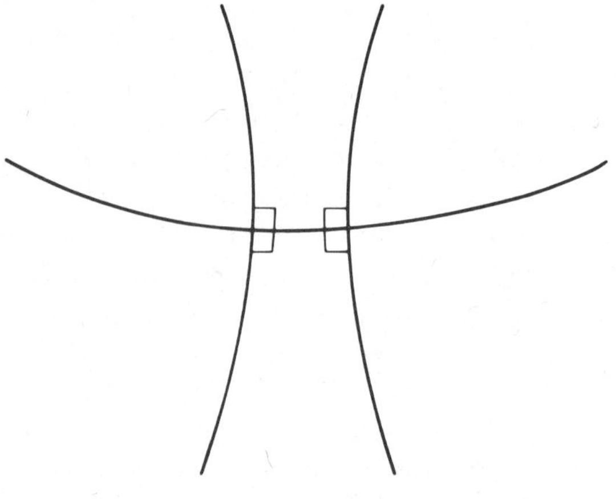

THEOREM L2 | If two lines are perpendicular to the same line, then they are nonintersecting and for any two nonintersecting lines there is exactly one line perpendicular to them both.

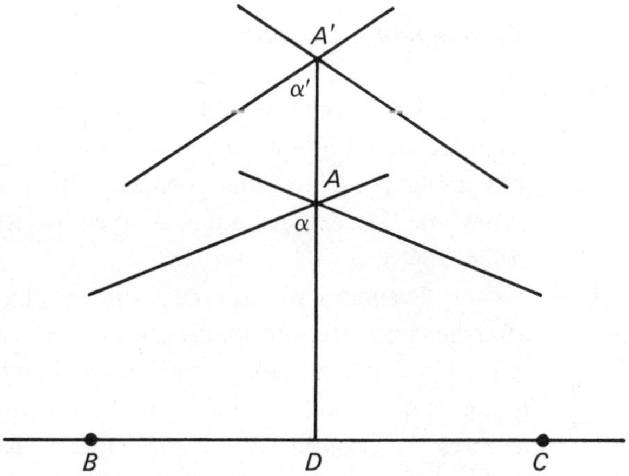

THEOREM L3 | The angle of parallelism at A for BC depends only on the distance of A from BC (the length of AD). This angle decreases as the distance increases and as the distance becomes small, the angle becomes very close to a right angle.

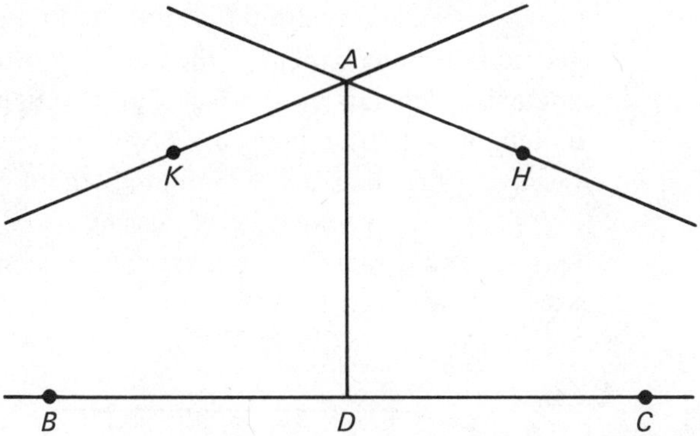

THEOREM L4 | If AH is the directed parallel to $\overrightarrow{BC}$, then the distance between AH and BC at a point X on AH decreases toward zero and becomes infinitesimally small as the point X moves in the direction of parallelism. The distance increases without bound as the point moves in the other direction.

Lobatchevsky developed this geometry by proving a large number of theorems about plane and three-dimensional figures. What he created is an axiom system that is as beautiful, complete, and valid as the system defined by Euclid.

Riemannian Geometry

The geometry of Bolyai and Lobatchevsky was at first generally regarded as unimportant mathematics. It was nearly forty years before it was understood how this geometry fits into the structure of mathematics as a whole. The explanation was given by Georg Friedrich Bernard Riemann (1826–1866).

Riemann was the best student that Gauss ever had. A few years after he completed his studies, Riemann was allowed to become a lecturer at the University of Göttingen. This was an unpaid position but carried with it the right to collect fees from those students who wished to attend the lectures. As a condition of "employment" Riemann was required to present a paper to the faculty. He chose as his topic, "The Hypotheses Which Are the Basis of Geometry." In this paper Riemann introduced the basic ideas and lines of development for many new geometries. (As we shall see later, the time had arrived when it became difficult to decide just what kinds of mathematical axiom systems should be called geometries.) We shall here consider only one of the many Riemannian geometries.

Riemann realized that the fact that parallel lines exist in Euclidean geometry is independent of the parallel postulate. The proof that a parallel line exists depends on the fact that a straight line can be extended so that its length is greater than any given number. This fact is often expressed by saying that lines have "infinite" length.

In the geometries of Saccheri and Lobatchevsky the distinctive feature is that there exists more than one such parallel line. In Saccheri's work with absolute geometry, the exclusion of the possibility of no parallels

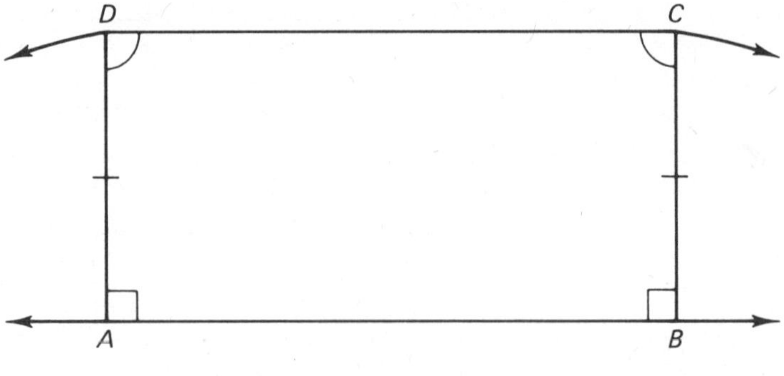

was accomplished by his proof that the Obtuse Angle Hypothesis about the summit angles of the quadrilateral leads to a contradiction. Saccheri showed that from this hypothesis it can be proved that the line through *CD* must intersect the line through *AB* *if* they are extended far enough.

Riemann noted, however, that Euclid's second postulate need not be interpreted to imply this. It is possible that a line be boundless in the sense that it "goes on forever" and has no ends without the line ever "going anywhere." This is possible when a line, such as a circle, doubles back on itself. He pointed out that if the second Euclidean postulate is interpreted in this weaker sense, so that it implies less, then there is a second alternative to Euclid's fifth postulate that will result in a consistent system.

Riemann investigated the geometry that results when this interpretation of the second postulate is made and the following equivalent of Saccheri's Obtuse Angle Hypothesis is added to the axioms of absolute geometry.

Riemann's Postulate: Any two straight lines intersect.

Riemann's development of this geometry was quite different from the approach used by Saccheri, whose work had been lost and was unknown at the time. Riemann's reasoning and ideas were more closely related to those of his mentor, Gauss. Some of the elementary results of this Riemannian geometry follow. Notice that in this geometry we can assume as a starting point for our list of theorems only those of the absolute geometry that do not depend on the tacit assumption of infinite length.

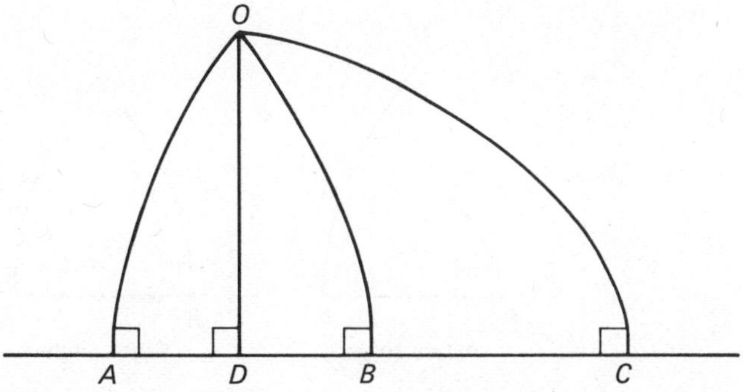

THEOREM R1 | All perpendiculars to a given line are concurrent (intersect at a point).

Proof: Let *AB* be a segment of the line *l*. Then the perpendiculars to *l* through *A* and *B* intersect at a point *O*. If a segment *BC* is laid off on *l* so that $|BC| = |AB|$, then the line through OC must be perpendicular to l (I-4). Also, if D is the midpoint of AB, then $OD \perp l$. The line from O to any point on l that can be constructed by repeated application of these procedures will clearly also be perpendicular. Since we can come as close as we please to any point on l with these procedures, we conclude that it is true for all points of l. (Here we have a vague argument, but it could be made precise if we were familiar with the ideas of limit and continuity that are basic to the careful study of mathematics. See Discussion Question 3.)

That is, not only do perpendiculars at all points of l pass through O, but all lines passing through O and intersecting l are perpendicular to l. In addition we see that if X and Y are on l, then $|OX| = |OY|$ because triangle OXY has two equal angles.

DEFINITION | A point through which all perpendiculars to a line pass is called a **pole** of the line.

DEFINITION | If l is a line, O is a pole for l, and A is on l, then $|OA| = d$ is called the **distance from the line to its pole.**

THEOREM R2 | The distance from a line to its pole is the same for all lines.

THEOREM R3 | In any right triangle ABC with right angle at B, the angle A is less than, equal to, or greater than a right angle as $|BC|$ is less than, equal to, or greater than d.

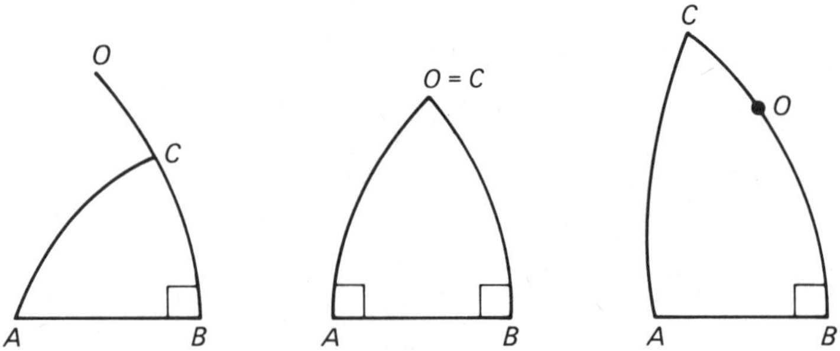

Acute angle   Right angle   Obtuse angle

Proof: If O is the pole to the line through AB, then O lies on the line through BC since angle B is a right angle. Therefore, $|OB| = d$. Draw AO. Then $\angle BAO$ is a right angle. If $|BO| > |BC|$, then $|\angle A| < |\angle BAO|$. That is, $\angle A$ is less than a right angle. The rest of the theorem follows similarly.

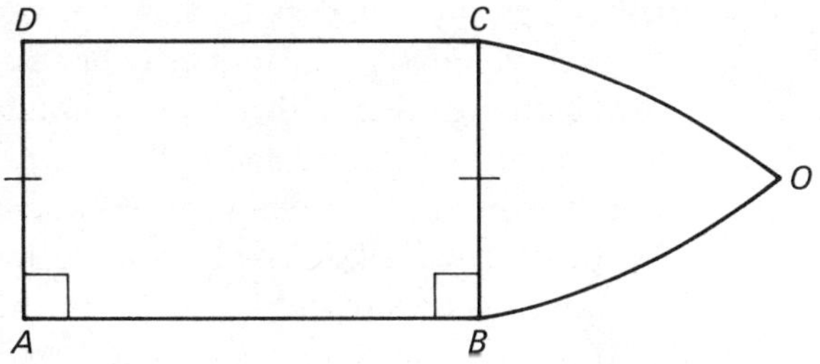

THEOREM R4 | The summit angles in a Saccheri quadrilateral are equal and obtuse.

Proof: The theorems about Saccheri quadrilaterals in absolute geometry, which were proved in Chapter 4, did not use the assumption of infinite length. Thus the angles C and D are equal and if E and F are the midpoints of AB and CD, then $AB \perp EF \perp CD$. If EB and FC are extended, then they must intersect at O, a pole for the line through EF. Since $|EO| = d$, we have $|BO| < d$. By considering triangle BCO we see that $\angle BCO$ is acute. It follows that $\angle BCF$ is obtuse and the theorem is proved.

This geometry has useful interpretations for "real-life" situations and its development for three-dimensional space and the study of the measurement of angles (trigonometry) and area is very interesting. These topics must, however, be left as outside reading for those interested.

Summary

Gauss, Bolyai, and Lobatchevsky each discovered that a consistent geometry could be developed by replacing Euclid's parallel postulate with a contradictory statement. The axioms that they chose are equivalent to the Acute Angle Hypothesis of Saccheri and the system is called Lobatchevskian Geometry. A generation later Riemann developed other geometries and showed, in particular, that if Euclid's second postulate is interpreted in an appropriate way, then a consistent geometry can be obtained in which parallel lines do not exist.

The creation of the new geometries is regarded as one of the greatest intellectual achievements of all time. Until the nineteenth century the postulates of Euclid were regarded as the only possible axioms since "they describe the way things really are in nature." With the discovery of the independence of the parallel postulate, mathematicians created new geometries as logically sound as Euclid's but very different from it. This led to the conclusion that geometry need not describe reality and posed the problem of deciding which of these sets of postulates, if any, does describe reality. This problem is unsolved and is likely to remain so for some time. Attempts to find the answer by experiment seem bound to fail because interpretation of the data obtained is based upon unsupportable assumptions about the physical properties of our universe. Furthermore, it may well be the case that there is no single geometry that describes all aspects of the real world. (See Discussion Question 1.)

The theorems of a geometry no longer express absolute truths: All that is known is that they are true if the postulates are true. It was this realization that led mathematicians to begin the study of different sets of postulates and their properties. The acceptance of Euclid's method of organizing mathematics as a postulational system and the discovery made more than two thousand years later that different sets of postulates can lead to distinct but equally consistent mathematical systems are jointly responsible for many of the great advances made in mathematics in the last one hundred years.

EXERCISES

1. Make a chart showing the Lobatchevskian and Riemannian analogues of the following Euclidean results.

(a) The Playfair Axiom.
(b) The sum of the angles of a triangle is two right angles.
(c) The sum of the angles of a quadrilateral is four right angles.
(d) The angle inscribed in a semicircle is a right angle.
(e) Squares exist.

***2.** Prove that in Riemann's geometry the sum of the angles in a right triangle is greater than two right angles. (*Hint*. Use Theorem R4, and follow Saccheri's proof of the analogous result on page 37).

3. Use Exercise 2 to prove that in Riemann's geometry the following two statements are theorems.

I. In any triangle the sum of the angles is greater than two right angles.

II. In any quadrilateral the sum of the angles is greater than four right angles.

4. Compare the postulates and theorems of Riemann's geometry with what you would want to be true for geometry on a sphere.

DISCUSSION QUESTIONS

1. Gauss attempted to determine which geometry describes the real world by measuring the angles of a huge triangle whose vertices were three mountains. His results were inconclusive. What kinds of experimental data *would have* enabled him to conclude that the real world is a model of one of the geometries?

2. What meaning or truth is there in the statement that "The shortest distance between two points is a straight line"? In particular, what will a "straight line" look like on various surfaces if this statement is used as the definition of a straight line?

***3.** Investigate the assumption made in this section that any point on the line could be approximated by an appropriate sequence of halfing and doubling operations starting with a given fixed length. (*Hint*. Call the given length 1 and see what numbers you can construct.) Think about variations of this procedure, such as using triples and thirds to approximate lengths.

six

the study of perspective

During the centuries after Euclid there were many developmen in geometry that were caused by the need for practical techniques to facil tate the solution of real problems. Now that we are familiar with the basi facts of Euclidean geometry and some other geometries that resulted fro the study of its properties as a formal system, we shall look at a ne geometry that was invented and studied because there was an immedia need to know the facts that its theorems contain. The development of th new geometry can be divided rather roughly into three parts. This wi emphasize a pattern found in the growth of many branches of mathematic The first part is the subject of this chapter and the next two parts a discussed in the one following.

Geometry in Renaissance Art

The artists of the early Renaissance began the systematic study perspective. The painters of this time were making the first attempts realistically representing three-dimensional objects on two-dimension media. They wished to develop general rules that would ensure that the pictures would accurately depict the actual subjects.

The man who is usually credited with being the first to develo a systematic method for doing this is Bruneleschi (1377–1446). The succe of Bruneleschi's method encouraged many others to investigate the matt and show their mastery of the new techniques in their works. Leon Battis Alberti (1404–1472), an artist and master architect, wrote a series of boo explaining his procedures. Leonardo da Vinci (1452–1519) wrote his *Treatise on Painting* which stresses the scientific method in painting and the importance of knowing mathematical principles. Albrecht Dürer (1471–1528) traveled to Italy to learn the secrets of perspective. He later improved upon the existing methods and published them in a book on geometry. All of these men were of necessity amateur mathematicians and there is some indication that Dürer studied geometry partly for the pleasure and intellectual stimulation it gave him. Each of them used a "cookbook" approach in which a specific type of problem is posed and the steps are given that must be followed to solve the problem.

The basic idea employed by these men is that if a plane, like a piece of glass, is placed between the artist and the scene that is to be depicted, then the image of the scene on the plane appears the same to the artist as the scene itself. If a collection of straight lines is drawn from points of the scene to the eye of the artist (a projection), then the image obtained by taking the points of intersection of the lines with the plane (a section) will accurately depict the scene.

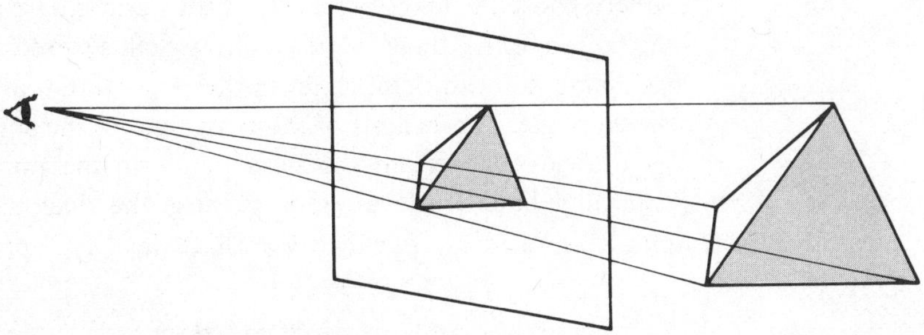

In theory it is necessary to draw a line from *each* visible point to the eye and this will result in *every* point of the painting being the image of *exactly one* such visible point. Such a mapping is called a *one-to-one correspondence* or a *one-to-one and onto mapping*. In practice only enough lines were drawn to establish a network within which the scene could then be sketched.

The instructions given by Alberti in his book *On Painting* in 1436 show that the artists knew and used theorems of elementary geometry as basic guides in their methods. Alberti first decided where, in relation to the scene, his rectangular "window" of canvas and the eye of the artist were to be supposed. That is, he located the point through which all the lines of vision had to pass and chose a section. He then put a point on the canvas at the foot of the perpendicular from the "eye" to the canvas. This point is the principal "vanishing point" and the line through it parallel to th bottom of the canvas served as the horizon or "vanishing line." He measure from the vanishing point a distance on the vanishing line equal to th supposed distance of the "eye" from the image plane. With these point of reference it is possible to establish an accurate image of a rectangula grid or checkerboard in the ground plane of the original scene.

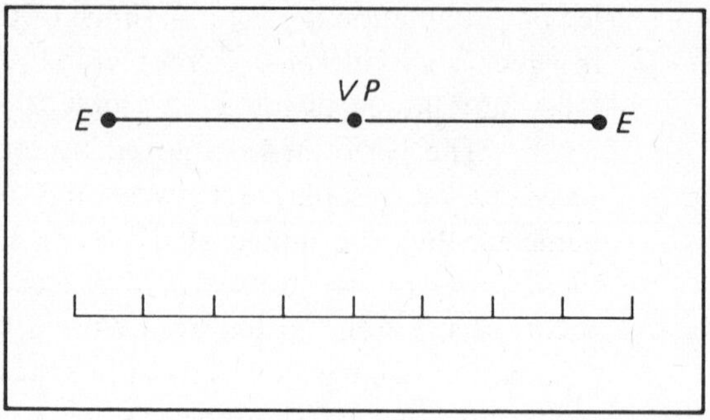

Alberti chose the length of a side of the basic square and laid this out alon the bottom of his image. The resulting points were then connected with th vanishing point to depict lines in the original that are perpendicular to th image plane. The heights of the cross lines of the squares were determine by drawing lines from the base points to the point constructed on th vanishing line. This was done because the diagonals of the squares ar parallel and must appear to vanish at this "eye" point.

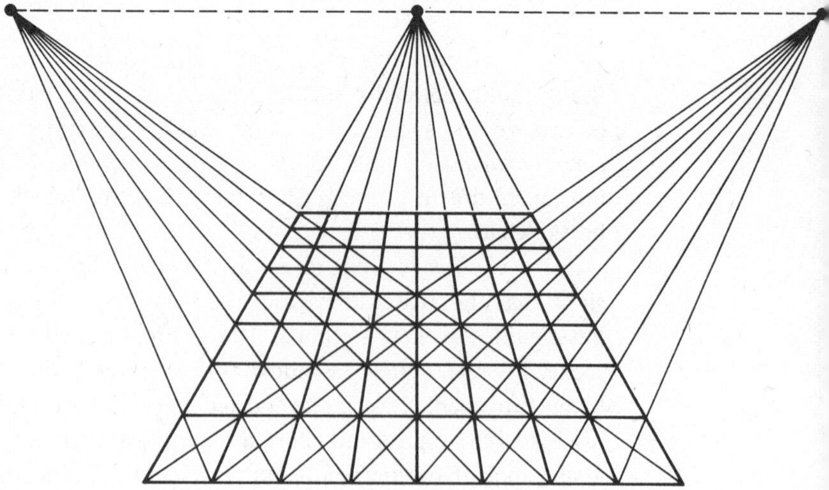

Perspective by Rodler. *An illustration from an early book* (*1531*) *on the techniques of perspective by Hieronymus Rodler. The vanishing point has been left in and labeled for the reader.* (*George Arents Research Library*)

Metric and Descriptive Properties

The artists and architects developed sophisticated techniques ar mechanical devices. Our interest lies, however, in the elementary facts th are apparent about the relationship between the original object and i image.

1. Length is usually changed in a projection. The size of the ima is not the size of the original and two identical lengths on the object m be different in the image. The amount of magnification depends on t location of the section between the point of perspective and the object. T object and its image are not congruent and congruent parts of the obje may have noncongruent images.

2. The image of any straight line is a straight line. In other word a projection preserves the property of *collinearity* of points.

3. Lines that intersect in the original have intersecting images. other words, a projection preserves the property of *concurrence* of straig lines.

4. Lines in the ground plane of the scene that are parallel to t section have parallel images. Vertical lines in the scene would also ha parallel images but parallels in any other direction will not have parall images. In fact, such parallel lines seem to approach each other and wou intersect if extended far enough.

5. Angles are changed. The image of a triangle is a triangle b it may well not be similar to the original. The image of a square is quadrilateral but probably not a square.

6. The image of a figure usually has an area unequal to that the original.

7. Circles have images that are not circles; however, the image a circle is usually an ellipse.

There are some general observations that can be made. Properti that depend on distance are called **metric** properties. They are usually n preserved by projections. Examples of these are length, area, parallelis and the size of angles.

Because collinearity and concurrence are preserved by projection certain other properties must also be retained by an image. The image a polygon must always be another polygon with the same number of side Properties that concern only the way the geometric objects are connect are called **descriptive** properties. The study of these properties is often call *descriptive geometry*. Notice that the word *geometry* is being used in a ne and in some sense self-contradictory manner. The name *descriptive geometry* here designates the study of a special set of results within Euclidean geometry.

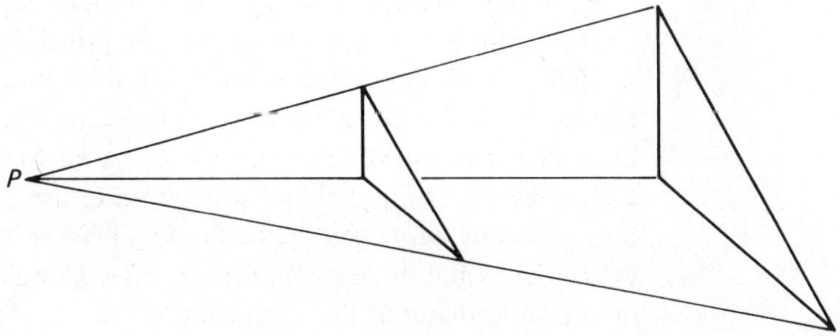

Certain special projections will preserve some metric properties. Consider a projection of an object that lies in a plane parallel to the section plane. In this case angles are not changed by the mapping. The image is similar to the original.

The farther away the point of perspective is chosen, the less that the lengths involved are changed. If the point of perspective could be chosen as a "point at infinity," then the lines emanating from it would be parallel and the image would be congruent to the original.

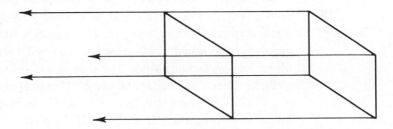

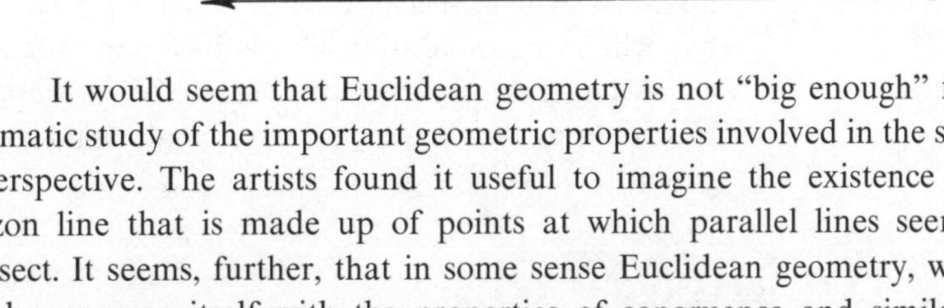

It would seem that Euclidean geometry is not "big enough" for a systematic study of the important geometric properties involved in the study of perspective. The artists found it useful to imagine the existence of a horizon line that is made up of points at which parallel lines seem to intersect. It seems, further, that in some sense Euclidean geometry, which mainly concerns itself with the properties of congruence and similarity, should be considered as a special case of this larger subject that is concerned with all projections.

Desargues and Pascal

One of the first men to systematically study the mathematical relationships involved in the theory of perspective was Girard Desargues (1593–1662). Desargues was an architect and engineer who wished for a clearer, more complete, knowledge of the principles involved. He published in 1639 a book entitled (approximately) "A Rough Draft of an Attempt to Deal with the Outcome of the Intersection of a Cone with a Plane." Desargues was unfortunately a very strong-minded, perhaps perverse is not too strong a word, individual who chose to use botanical terminology for his definitions. Moreover, he had only a few copies printed and gave them to his friends and students. As a result, his work was unknown to professional mathematicians until the nineteenth century.

Another man who contributed to the development of the mathematical theory of perspective in the seventeenth century was Blaise Pascal (1623–1662). As a young man Pascal was impressed with Desargues' work and proved several beautiful theorems, one of which bears his name and is discussed in the next chapter. Unfortunately, much of Pascal's work was not published. All copies of his principal paper on projective geometry have been lost for more than two hundred years and we have only secondhand information about its content.

Pascal had wide interests as well as great talent. He made several important discoveries in mathematics and physics. He is now considered to be one of the first contributors to existential philosophy. Probably one of the reasons that Pascal embraced Desargues' geometry is that most mathematicians of this time ignored Desargues in order to study the analytic geometry, which had just been introduced by Descartes as an appendix to his major philosophical work. Pascal disagreed with the religious and philosophical views of Descartes and chose not to work on the analytic geometry.

Pascal's contributions to mathematics were not as numerous as they might have been because he suffered from chronic ill health, as well as hypochondria, and during his relatively short life had several periods during which his religous fervor caused him to give up mathematics completely. It was almost two centuries before mathematicians began a serious investigation of projective geometry.

Summary

In the fourteenth and fifteenth centuries artists began to apply basic geometrical facts in attempts to achieve realism in their work. Various procedures were developed that became the method of perspective. During the seventeenth century amateur mathematicians such as Desargues and Pascal contributed to the mathematical theory of the technique.

It was seen that the metric properties of geometric figures are not necessarily preserved in their images under a projection. Descriptive properties, however, are invariant under such mappings. The extension of Euclidean geometry that is necessary to properly study projections and their effects is called *projective geometry*.

EXERCISES

1. Determine whether or not the image of each of the following kinds of geometric figures is a figure of the same kind.

(a) Triangle.
(b) Isosceles triangle.
(c) Equilateral triangle.
(d) Square.
(e) Parallelogram.
(f) Quadrilateral.

2. Explain why the property of a curve being a circle is a metric property.

DISCUSSION QUESTION

Choose a point in the plane as the point of perspective. Draw a line, and label some points (say 3 or 4) on it. This will be the domain or "object." Draw another line, between the domain line and the point, to serve as the image line. Connect the point of perspective with each of the points on the original line to locate their images. Such a mapping is called a *perspectivity*.

(a) Is the distance between points preserved under a perspectivity? Is it possible to choose the point of perspective and the section line so that distance is preserved?

(b) Check to see that a perspectivity is a one-to-one correspondence between the two sets of points. Will it always be a one-to-one correspondence no matter how many points are put on the original line? (When a mapping is a one-to-one correspondence and a particular set is chosen as the domain and the other as image, then the mapping in which they are chosen in the other order is called the *inverse mapping*.) What interpretation should be given the point of intersection of the domain and image lines (if they intersect)?

(c) If given two lines with points A, B, C on one and A', B', C' on the other, what must be true if the latter are to be the images of the former under a perspectivity?

(d) Check out the following procedure that purports to show that it is always possible to map any three collinear points into any three others if two perspectivities are combined (done consecutively). Given two lines: l_1 with points A_1, B_1, C_1 and l_2 with points A_2, B_2, C_2. The first point of perspective will be C_2. The image line is determined as the line joining the two points of intersection: A_1C_2 with A_2C_1 and B_1C_2 with B_2C_1. Label the images of A_1, B_1, C_1 under this perspectivity as A', B', and C'. Now if C_1 is chosen as the second point of perspective and l_2 as the image line, then the points A', B', C' are mapped to A_2, B_2, C_2.

(i) Does this procedure always seem to work no matter what lines are drawn and where the points are placed on the lines?

(ii) Would the procedure work if A_1 and A_2 or B_1 and B_2 were used as the points of perspective instead of C_1 and C_2? Do you get a different intermediate image line when different pairs are used as points of perspective?

seven

projective geometry

In order to continue the story of the development of projective geometry, it is necessary to jump forward temporarily to the time of the French Revolution. During the seventeenth and eighteenth centuries very few mathematicians studied the descriptive properties of geometric figures. As discussed in the next chapter, powerful new geometric techniques were invented in the seventeenth century and most mathematicians believed that these methods were more important than the study of perspective because of their wider applicability to practical problems.

Gaspard Monge (1746–1818), like most French mathematicians of his time, had close ties with the army. The best education in mathematics could be obtained in the schools for military officers, and many prominent mathematicians were supported by these institutions. In this time of political upheaval and the subsequent adventures of Napoleon, many mathematicians also held high government or military positions. Monge was instrumental in the founding of a school for military engineers and was probably its greatest teacher. He developed the study of the descriptive properties of three-dimensional objects and taught the techniques of representing such objects by their projections onto planes. The course he taught was more similar to a modern course in mechanical drawing than to a course in mathematics. Monge had many students who became famous mathematicians.

Jean-Victor Poncelet (1788–1867) was a student of Monge. He finished his studies just before Napoleon invaded Russia. Poncelet was captured in Russia and spent a few years as a prisoner of war. While a prisoner he conceived his *Treatise on the Projective Properties of Figures,* which was published in 1822 after his return to France. Poncelet's wo marks the beginning of the second period in the development of projecti geometry. Prior to Poncelet the emphasis had been on discovering the rul of perspective that were necessary to accurately depict objects. The fe mathematicians who had worked in the field had concerned themselves wi finding the methods to solve specific problems. Poncelet began the syster atic study of the geometrical properties that are preserved by projecti mappings.

Theorems from Projective Geometry

As seen in the discussion of perspective, it is inconvenient to stuc the properties of projections within the framework of Euclidean geometr It is useful to enlarge the set of lines and points in Euclidean geometry b introducing an "imaginary" or "ideal" line made up of points at whic parallel lines intersect. Poncelet worked with Euclidean geometry and use all of Euclid's postulates. In order to avoid having to continually state th exceptions to his theorems, which arise when some of the lines involve do not intersect, he made use of the "extra" line of points of intersectic for parallel lines.

It is useful to consider what kinds of facts the theorems of projecti geometry can be expected to contain. The theorems that we have seen i the geometries of Euclid, Lobatchevsky, and Riemann dealt with conditio under which figures are congruent or similar and provided information abo the area and angle size. We expect that the theorems in projective geomet should give facts about the relationship of a figure to its image und projection. Metric properties are not preserved by most projections. The onl properties that we found to be unchanged by all such mappings were thos that depend on the preservation of collinearity and concurrence.

With this background the subject matter of the following definitio and theorems from projective geometry should not be too surprising.

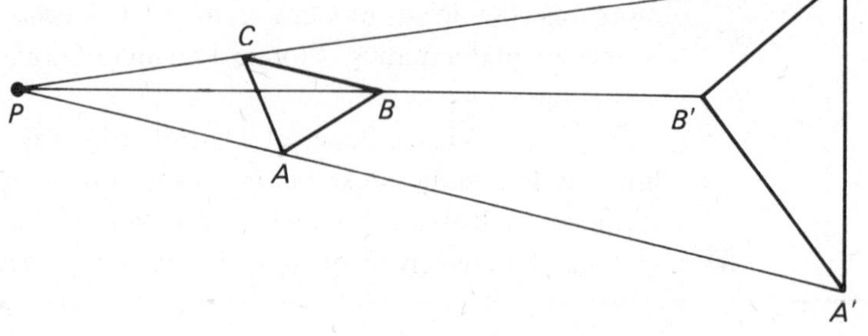

DEFINITION | Two triangles are said to be **perspective from the point *P*** if there is a one-to-one correspondence between the vertices of the triangles in such a way that the lines joining corresponding vertices intersect at *P*.

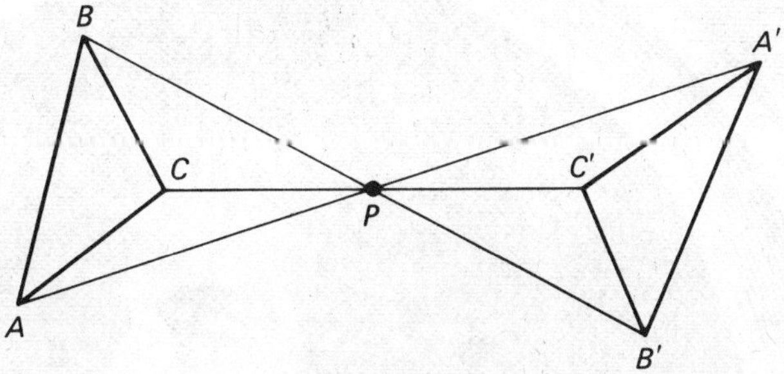

One of the basic results of projective geometry is a theorem originally proved by Desargues. When the theorem is stated in terms of the definition above, the statement is simple and straightforward. (An even more elegant statement of the theorem is asked for in Discussion Question 2.)

DESARGUES′ THEOREM | If two triangles are perspective from a point *P*, then the corresponding pairs of extended sides intersect in points which are collinear.

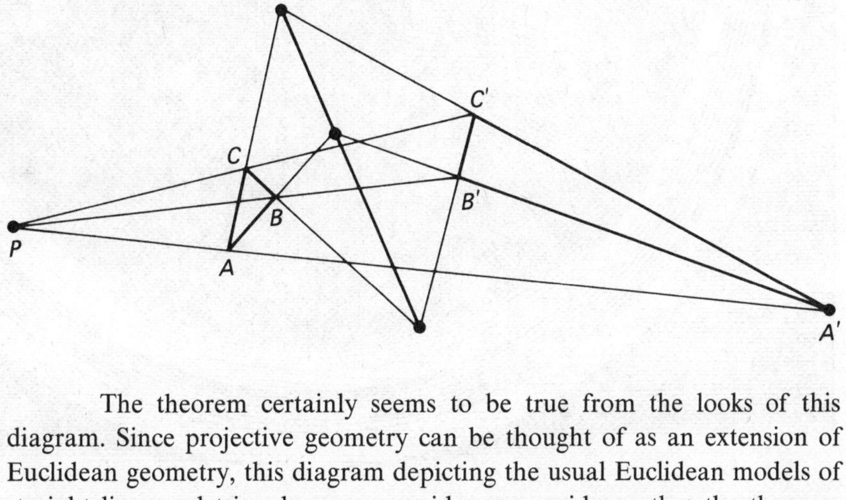

The theorem certainly seems to be true from the looks of this diagram. Since projective geometry can be thought of as an extension of Euclidean geometry, this diagram depicting the usual Euclidean models of straight lines and triangles may provide some evidence that the theorem is true. When the projective geometry is developed as a distinct axiomatic system, however, care must be taken to ensure that intuitive ideas about straight lines do not intrude.

Desargues' theorem should be true in Euclidean geometry as long

Blaise Pascal. (*George Arents Research Library*)

as parallel lines are not involved. It is easy to see that difficulties arise when a pair of corresponding sides of the triangles do not intersect. (See Exercise 3.) One of the axioms of projective geometry excludes the existence of parallel lines.

Another famous theorem of projective geometry was first published as a one-page leaflet by Pascal in 1640. Although much of the beauty of this theorem lies in its great generality, we shall restrict ourselves here to a special case. (See p. 80 for a more general statement.)

PASCAL'S THEOREM If each set of three alternate vertices of a hexagon lies on a straight line, then the three points of intersection of pairs of opposite sides are collinear.

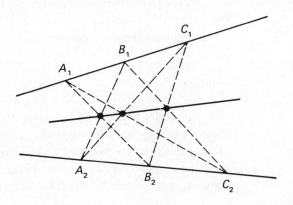

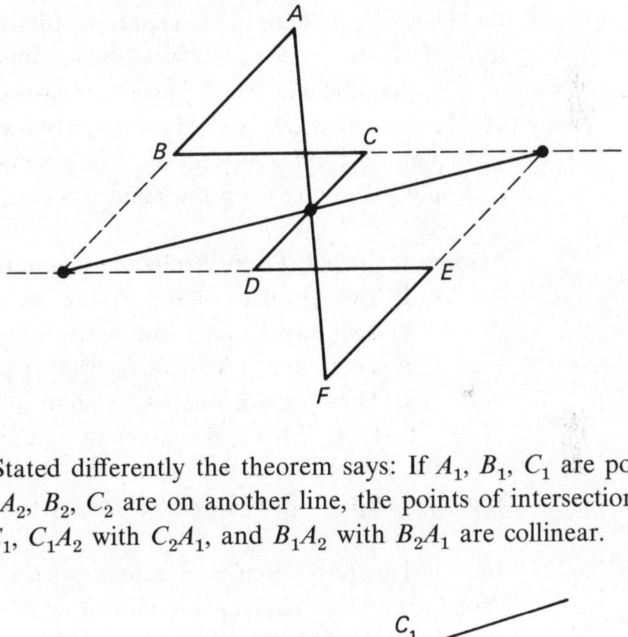

Stated differently the theorem says: If A_1, B_1, C_1 are points on a line and A_2, B_2, C_2 are on another line, the points of intersection of B_1C_2 with B_2C_1, C_1A_2 with C_2A_1, and B_1A_2 with B_2A_1 are collinear.

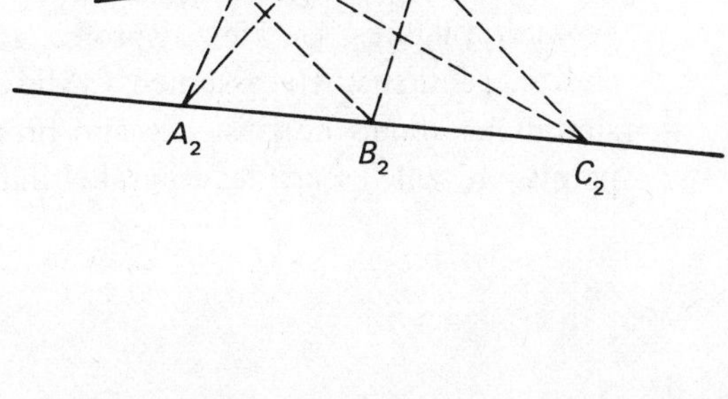

This theorem also seems to be true in Euclidean geometry provided car is taken to avoid situations in which parallel lines occur. It was in fact proved with these exceptions, by Pappus about 300 A.D.

Axioms for Projective Geometry

Until the end of the nineteenth century, projective geometry wa viewed as an extension of Euclidean geometry rather than a separate geom etry. After the discovery and acceptance of the non-Euclidean geometries it was realized that projective geometry could and should be developed a an independent axiomatic system. When this was done, it was seen that th basic structure of projective geometry does not depend on the concept o Euclidean parallel lines and the interrelationships among projective geome try and all the other geometries became much clearer.

To conclude the discussion of projective geometry, we shall list a set of axioms that can be used for projective geometry. The undefined term will again be *point, line,* and *on.* It is assumed that a point is on a line i and only if the line is on the point.

Axioms for (Real) Plane Projective Geometry

1. Two distinct points are both on exactly one line.
2. Two distinct lines are both on exactly one point.
3. There are at least three distinct points on each line.
4. There exist a line and a point not on the line.
5. If A_1, B_1, C_1 are points on one line and A_2, B_2, C_2 are points on a distinct line, then the three points that are on A_1B_2 and A_2B_1, A_1C_2 and A_2C_1, B_1C_2 and B_2C_1 are all on the same line.

This last axiom is of course essentially the theorem attributed to Pascal. There are equivalent axioms that could be used in its place but a consideration of these statements would lead us too far afield.

Summary

Poncelet began the systematic study of the projective properties of geometric figures. He viewed projective geometry as an extension of Euclidean geometry. He assumed Euclid's axioms and made additional assumptions about the existence and properties of an "ideal line" of "ideal points" at which Euclidean parallel lines intersect.

About 1900, projective geometry was developed as a formal axiomatic system independent of the concepts of Euclidean geometry. The study of projective geometry then became the investigation of an abstract mathematical structure and emphasis was put on determining its relationships with other geometries.

EXERCISES

1. Experiment with different placements of the triangles mentioned in Desargues' theorem, and check the conclusion graphically.

2. You will notice that there is no mention made in Pascal's theorem about the order in which the points involved are placed on the lines or the appearance of the hexagon. (After all, hexagon only means *six-sided figure*.) Verify the conclusion of the theorem for a variety of the possible configurations.

3. Modify the statements of Desargues' and Pascal's theorems so that they become correct theorems in Euclidean geometry. Be explicit about what happens when one or more pairs of corresponding sides are parallel.

DISCUSSION QUESTIONS

1. Check the following procedure for constructing the line through a given point P and an inaccessible point Q that is the point of intersection of two given lines, l_1, l_2. Choose A_1 and B_1 on l_1 and A_2 and B_2 on l_2. Let C_1 be the intersection of A_1B_2 with B_1P, C_2 be the intersection of B_1A_2 with B_2P, and let Q' be the intersection of A_2C_1 and A_1C_2; then PQ' is the desired line. Can you think of any practical applications for this technique?

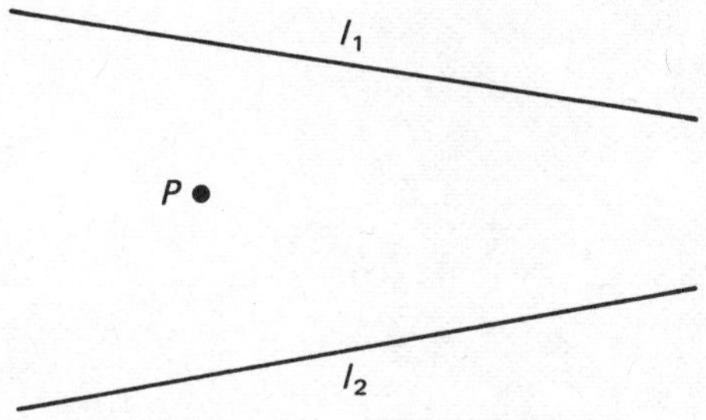

***2.** One of the important facts about projective geometry is that the Principle of Duality is true. This result says that if the words *point* and *line* are interchanged in any theorem (and appropriate changes are made in the phrasing), then the resulting statement is also a theorem. For instance, the first two axioms for projective geometry are "duals" of each other.

(a) Find the dual statements of the remaining three axioms, and draw diagrams to illustrate these results.

(b) Define the concept of "perspective from a line *l*" by stating the dual of the definition of perspective from a point.

(c) Restate Desargues' theorem in terms of this definition.

(d) State and illustrate the dual of Desargues' theorem.

eight

analytic geometry

Alexandria remained the center of scientific activity for several hundred years after the death of Alexander. A succession of great Greek mathematicians lived there and added to the body of mathematical knowledge. The study of science gradually declined, however, and the nature of scientific inquiry changed when the Romans conquered the Mediterranean region.

The Romans were interested in only those parts of science that could be applied to solve practical problems. Little progress was made in the development of mathematics as an abstract logical subject during the time of the Roman Empire. Civilization in Europe was completely disorganized after the fall of the Roman Empire and the knowledge of the Greeks survived only in the monasteries. It is the Arabs and the Byzantine Greeks who are primarily responsible for the preservation and transmission of the Greek classics.

The first significant revival of mathematics as a creative science was in the fifteenth century. By this time the Black Death had passed, universities had been founded in the major cities, and the printing press had been invented. The Renaissance had begun. It was a period of major achievements in art and the sciences. The origins of a large number of important scientific disciplines are to be found in this period.

By the beginning of the seventeenth century, a need for a great deal of practical geometry had arisen. The advances made by Kepler in astronomy, the need for navigational aids in sailing the open seas, the use of artillery in war, the utilization of lenses to focus light, and the work of Galileo and Newton on the theory of the force of gravity all made it important that a practical method for computing the numerical values related to geometric figures be developed.

René Descartes (1596–1650) and Pierre de Fermat (1601–1665) are both credited with contributing essential innovations in geometry that have since been developed into the branch of mathematics called *analytic geometry*. Descartes was born into a noble French family and grew up during the time celebrated by Dumas in his *The Three Musketeers*. As a young man Descartes lived a life befitting his position, but he soon turned to a life of seclusion and meditation. His major interests were philosophy and science and in 1637 he published his "Discourse on the Method of Reasoning Well and Seeking Truth in the Sciences." This book contained an appendix on geometry in which Descartes presented the new techniques he had developed.

Fermat was a royal magistrate whose avocations were mathematics, literature, and verse. He was particularly interested in recreating some of the great Greek works on mathematics which are mentioned in various historical accounts but which had never been found. Very few of Fermat's mathematical results were published during his lifetime and he is perhaps most famous as a mathematician for something he probably didn't actually do.

Diophantus of Alexandria wrote his *Arithmetica* about 300 A.D. One of the problems in this book is that of finding triples of whole numbers a, b, c such that $a^2 + b^2 = c^2$. The existence of such sets, known as Pythagorean triples because of their relation to the sides of right triangles, has been known since the time of the Babylonians. The sets 3, 4, 5 and 5, 12, 13 are examples. In Fermat's copy of the *Arithmetica* there is a marginal note in which he claims to have found a proof that there are no triples of integers such that the sum of the cubes, or fourth powers, or any other higher power, of two of the numbers is equal to the same power of the third number. For more than three hundred years mathematicians have been trying to decide whether or not Fermat really proved this and, if he did, how he might have done it. No proof has ever been found for his statement that $a^n + b^n = c^n$ is impossible when $n > 2$. Many useful new techniques in algebra have been invented in the attempt to prove Fermat's Last Conjecture (or Fermat's Last Theorem, depending on your faith in the content of the marginal note).

Fermat and Descartes discovered that it is possible to combine the methods and concepts of geometry and algebra in a way that is useful in both branches of mathematics. Their basic idea was to associate algebraic equations in two unknowns with geometric curves in the plane. The geo-

René Descartes. *Descartes' goal was to unify science. After seeing the fate of men such as Galileo at the hands of the Inquisition, Descartes restricted his scientific publication to a description of the principles of reasoning. His work on analytic geometry was appended as an example of the application of his principles.* (*George Arents Research Library*)

metric properties of a curve then correspond to the algebraic properties of its associated equation, and conversely. Using this correspondence, problems in geometry can be translated into problems solvable by algebraic manipulation, and insight into the nature of solutions of equations can be gained by consideration of the geometric interpretations.

Both Fermat and Descartes got the inspiration for their methods from studying similar techniques in the works of the ancient Greek mathematicians. Fermat and Descartes had the advantage of being able to use algebraic methods and symbolism which were unknown to the Greeks, and, what is most important, their choice of the type of equations which are to be associated with the curves was much better than the choice made by the Greeks.

Thus analytic geometry was developed as a method for doing geometry and applying geometric techniques in other branches of mathematics. The results of analytic geometry were almost immediately used in a wide variety of applications. It is necessary to know some of the basics of analytic geometry if one is to understand the nature of much of the mathematics developed since the seventeenth century.

At the end of the nineteenth century it was realized that it is possible to use the concepts of analytic geometry to provide a model for Euclidean geometry. This was of some importance because the discovery of the non-Euclidean geometries made it desirable that Euclid's postulates be shown to be consistent. The manner in which some elementary facts about analytic geometry are introduced in what follows should indicate how a model for Euclidean plane geometry can be provided by interpreting the undefined terms as sets of numbers. (A logically rigorous discussion of these ideas would require that the numbers be defined, independently of any geometric concepts, and that the properties of the numbers be established. We shall investigate these matters in later chapters. At this point we shall simply assume that there are sufficient numbers, with "well-known" properties, to do what we wish to do.)

Points and Straight Lines

Although it is probable that the reader has had some experience with the use of Cartesian coordinates, let us review the basic ideas as they apply to Euclidean plane geometry. We choose a point O in the plane to be the origin. Through O we draw two perpendicular lines, which will be called the x-axis and the y-axis. Then we select a unit of length and a positive direction on each axis. Having done this we can describe every point in the plane by providing two numbers, the distance from the y axis and the distance from the x axis.

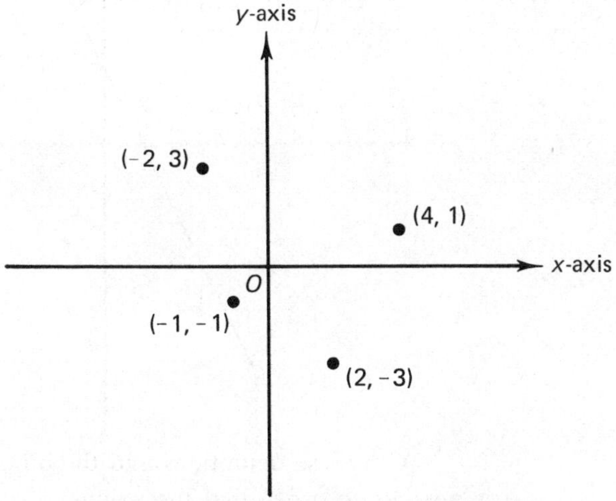

The graphs that we shall draw will provide extremely valuable guidance for our intuition but are *not* a necessary part of the formal development. In our considerations of mathematics we shall have more use for the guidance than the formal details.

If analytic geometry is to provide a model for Euclidean geometry, then it is necessary to give meaning to the undefined terms. A **point** is defined as an ordered pair of numbers. The numbers are called the coordinates of the point. The fact that the pairs of numbers are ordered means, for example, that (4, 3) and (3, 4) are different points.

A **straight line** is defined as a *set of points* (x, y) that satisfy an equation of the form $ax + by = c$ with a and b not both zero. Given any three numbers a, b, and c that satisfy the nonzero condition, there is exactly one line which consists of the points which satisfy the equation.

Example: The set of all points of the form (x, y) whose coordinates satisfy the equation $7x - 3y = 1$ is a straight line. The point (1, 2) is on this line since $7(1) - 3(2) = 1$. The point (4, 6) is not on the line since $7(4) - 3(6) \neq 1$.

Some additional examples are shown in the following diagram.

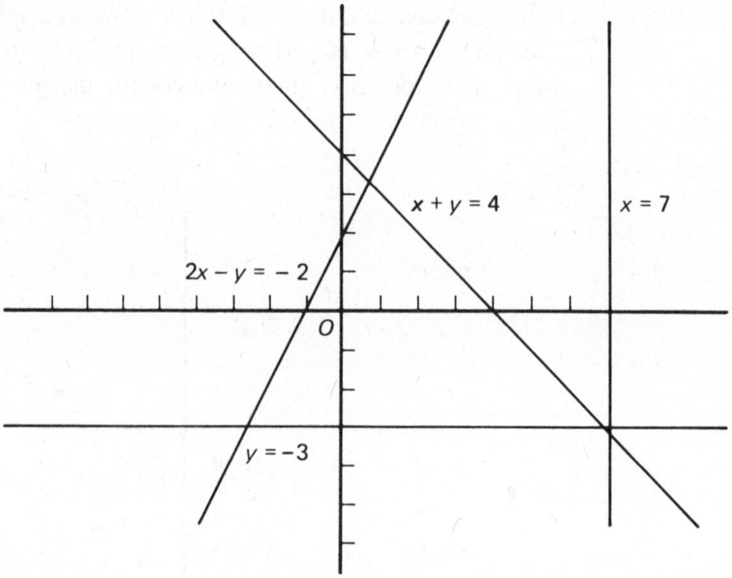

With these definitions and the others that are suggested by them, it is possible to prove that the axioms of Euclidean geometry are true in this model. In particular it is possible to show that any two points are on exactly one straight line. The proof that there is at least one straight line through two given points is constructive in the sense that it shows how to find the equation for the line. It will be useful to be able to do this.

Notice that if the two given points have identical first or second coordinates, then the line through them is easily found. For example, the line through (1, 4) and (7, 4) has equation $y = 4$ and the line through (6, -1) and (6, 43) has the equation $x = 6$.

DEFINITION | If b is not zero, then the **slope** of the line with equation $ax + by = c$ is defined to be the number $-a/b$. If $b = 0$, the slope of the line is not defined.

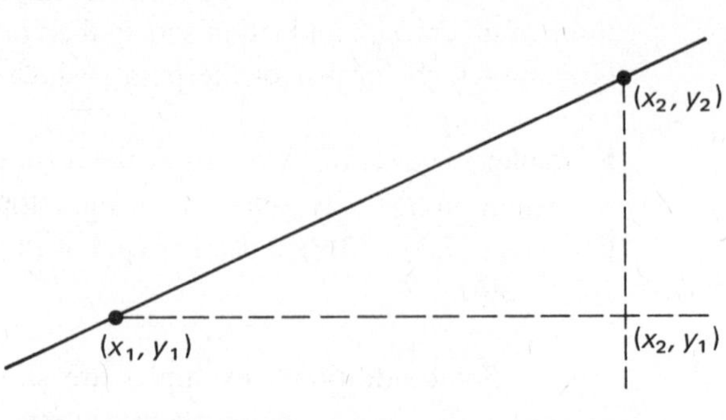

THEOREM | If (x_1, y_1) and (x_2, y_2) are points on a line, and $x_1 \neq x_2$, then the slope of the line is $(y_2 - y_1)/(x_2 - x_1)$.

If $ax + by = c$ is the equation of the line, then $ax_1 + by_1$ and $ax_2 + by_2$ must both equal c. Equating these two quantities gives $ax_1 + by_1 = ax_2 + by_2$ from which simple manipulations yield the desired result.

$$a(x_1 - x_2) = b(y_2 - y_1) \quad \text{or} \quad -\frac{a}{b} = \frac{y_2 - y_1}{x_2 - x_1}$$

This theorem can be used to find an equation for a line containing two given points (provided the points do not have the same first coordinates). If a point (x, y) is to be on the same line with the given points (x_1, y_1) and (x_2, y_2), then the slopes

$$\frac{y_1 - y}{x_1 - x} \qquad \frac{y_2 - y}{x_2 - x} \qquad \frac{y_2 - y_1}{x_2 - x_1}$$

must all be equal. The geometric interpretation of this fact is that the triangles shown below must be similar. Therefore, the ratios of the corresponding sides must be equal. Equating either of the first two ratios with the third will yield an equation for the line.

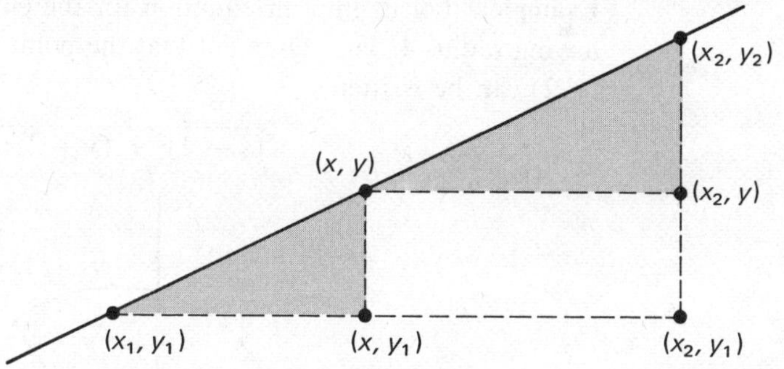

Example: A point (x, y) is on the line through (2, 3) and (5, 4) if and only if

$$\frac{y - 3}{x - 2} = \frac{4 - 3}{5 - 2} = \frac{1}{3}, \qquad 3(y - 3) = (x - 2), \qquad x - 3y = -7.$$

The answer is readily checked by substituting the coordinates of the two points into the equation. Check to see for yourself that the use of the other ratio yields the same equation.

Circles

Euclid's definition of a circle depends on the concept of distanc The appropriate definition of distance is suggested by the diagram belo and the fact that the Pythagorean rule for right triangles should hold.

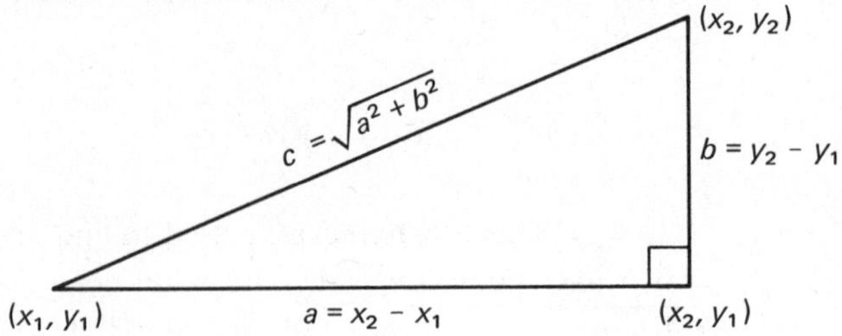

DEFINITION | The **distance** between the point (x_1, y_1) and the point (x_2, y_2) is the nor negative number

$$\sqrt{(x_2 - x_1)^2 + (y_2 - y_1)^2}.$$

DEFINITION | A **circle** is a set of points that are equidistant from some fixed point. Th fixed point is the **center** of the circle and the distance is the **radius** of th circle.

Example: Let us find an equation for the circle with center at (1, 2) an having radius 4. The statement that the point (x, y) is a distance of 4 fror (1, 2) can be written

$$\sqrt{(x - 1)^2 + (y - 2)^2} = 4.$$

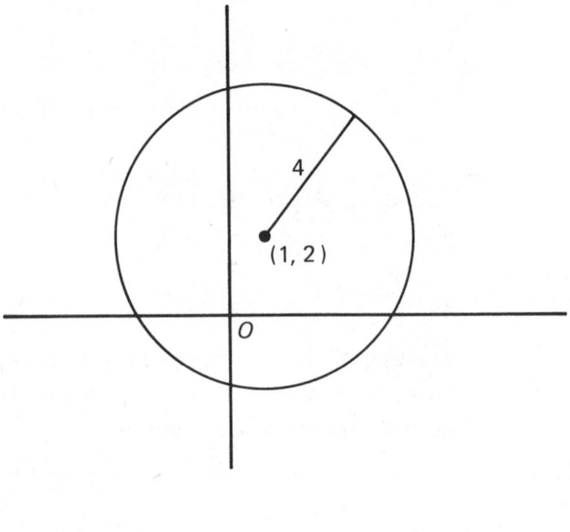

Thus the point (x, y) is on the circle if and only if its coordinates satisfy this equation. It is customary to simplify equations of this form by squaring both sides. The equation then becomes

$$(x - 1)^2 + (y - 2)^2 = 16.$$

From the definitions it follows that in general the equations of a circle with center at the point (a, b) and radius r are

$$\sqrt{(x - a)^2 + (y - b)^2} = r$$
$$(x - a)^2 + (y - b)^2 = r^2.$$

Summary

Descartes and Fermat discovered a very useful way of associating geometric curves with algebraic equations. The techniques that they originated have been developed into the subject now called *analytic geometry*. Thus analytic geometry is a collection of methods for investigating and applying existing geometries rather than a distinct axiomatic system.

The methods of analytic geometry can be applied to construct a model for Euclidean plane geometry. In this model Euclid's undefined terms are defined as sets of numbers. (This, of course, presupposes that the properties of sets of numbers have been established.) A point is defined as an ordered pair of numbers and a straight line is defined as a set of points that satisfy a special type of algebraic equation. After Euclid's undefined concept of distance has been defined, it is possible to find and characterize the equations satisfied by the points on circles.

EXERCISES

1. Verify each of the following statements.

(a) The points (1, 1) and (3, −4) are on the line $5x + 2y = 7$.

(b) An equation for the line through the points (6, 2) and (−5, −1) is $3x - 11y = -4$.

(c) The distance between (1, 2) and (7, 6) is $\sqrt{52}$.

(d) The distance between (1, −5) and (−2, 6) is $\sqrt{130}$.

(e) An equation for the circle with center at (1, −4) and radius 5 is $(x - 1)^2 + (y + 4)^2 = 25$.

(f) The points (−3, 7), (−3, 3), (−1, 5), and (−5, 5) lie on the circle $(x + 3)^2 + (y - 5)^2 = 4$.

2. Pick out a few specific points—say four or five—and plot them on a coordinate system. Calculate the distance between various pairs of points. Find the equations for the lines between pairs of points, and find the slopes of the lines. You may well decide at this stage to put in or take out some points to achieve variety. Remember that you can check your equations by substitution of the coordinates of the points. Continue until you are satisfied you have the ideas fixed.

3. Choose different specific points to be the centers of circles and different values for the radii. Find the equations of the circles. Graph the circles. You can check your results by making sure, by substitution, that the appropriate points, above, below, to the right, and to the left of the center are on the circle. For example, (2, 7), (2, −1), (6, 3), and (−2, 3) are easily seen to be on the circle $(x - 2)^2 + (y - 3)^2 = 16$, which has center (2, 3) and radius 4.

DISCUSSION QUESTIONS

1. (a) Find an equation which is satisfied by exactly those points which are equidistant from the points (1, 2) and (3, 4). Show graphically where these points are. What can be said in general about the set of points equidistant from two given points?

(b) Explain how two applications of the procedure in part a could be used to find the equation of a circle passing through three given points. Will this procedure always work? Under what circumstances can you find a circle that passes through four given points?

2. In order to complete the demonstration that Euclid's axioms are true in the model defined in this chapter it is necessary to find criteria for deciding whether two given lines are parallel or perpendicular.

(a) Find conditions on the coefficients a, b, d, and e that will be satisfied if the lines $ax + by = c$ and $dx + ey = f$ are parallel. Will the lines be parallel whenever these conditions are satisfied? (*Hint.* Use the idea of slope.)

(b) Repeat part a with the word parallel replaced by the word perpendicular.

*(c) Verify Euclid's fifth postulate in this model.

***3.** Show that each of the following theorems from Book I of the *Elements* is true in the model discussed in this chapter by using the methods of analytic geometry to prove them.

(a) I-10 (*Hint.* Find a formula for the midpoint of the line segment between two given points.)

(b) I-30

(c) I-11

(d) I-1

(e) I-31

(f) I-20

nine

the conic sections

There are curves other than straight lines and circles that continually arise in discussions of mathematics. Usually such a curve is of interest because there is an important geometrical relationship satisfied by the points on the curve. The curves that are the subject of this section were given their names by the Greeks who knew that these curves could be obtained by passing a plane through a right circular cone.

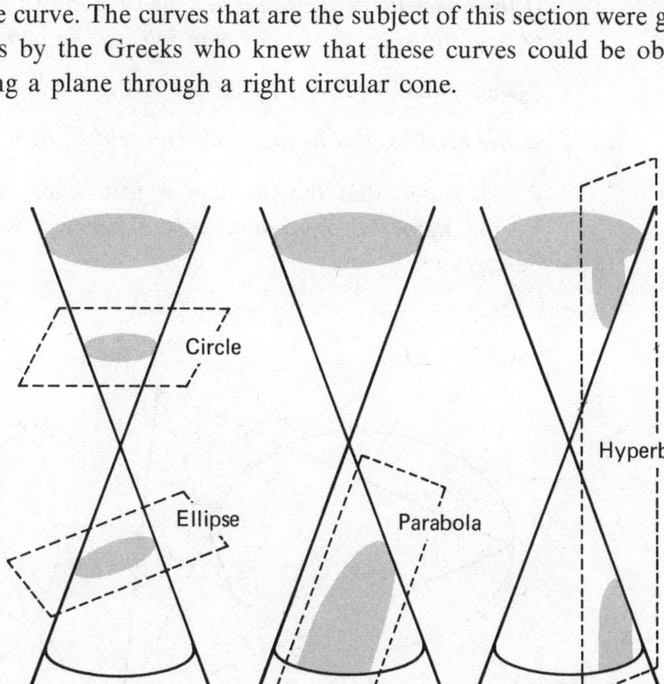

When the plane cuts entirely through one of the two nappes (ends) of the cone, the section is an ellipse. If the plane cuts only one nappe and does not cut completely through the cone, then the curve is a parabola. If the plane cuts both nappes, the section is an hyperbola. Circles and pairs of straight lines are also technically conic sections since they can be obtained as special cases of the ellipse and hyperbola.

The conic sections were studied extensively by the Greeks. The mos famous surviving work, *Conics,* was written by Apollonius about 250 B.C Apollonius studied mathematics in Alexandria shortly after the death of that school's most famous teacher, Euclid. Euclid himself apparently wrote a work on conic sections but no copy has ever been found. A revival of interest in these curves came in the seventeenth century. Desargues' investi gation of perspective dealt mainly with sections of cones. (See p. 58. Desargues and Pascal realized that the image of a conic section under projection is another conic section. They knew that any theorem which gives descriptive properties of a circle, for example, must also be true for a conic section which is the image of the circle under a projection.

The theorem attributed to Pascal in the section on projective geometry—

If the alternate vertices of a hexagon lie on two straight lines, then the pair of opposite sides of the hexagon intersect in points which are colinear,

—was actually proved by Pascal in the form,

If the vertices of a hexagon lie on a circle, then . . . .

Pascal stated that the theorem is true when the word *circle* is replaced by the name of any conic section because the properties involved ar descriptive.

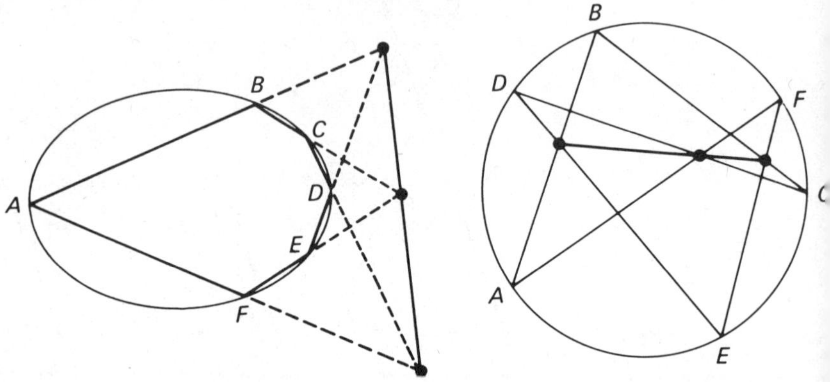

Both Descartes and Fermat were familiar with the work of Apollonius. In fact, some scholars think that Apollonius should receive a large share of the credit for originating the techniques of analytic geometry. Descartes continually mentioned in his works that his methods could be used to solve problems involving the conic sections that had been posed by the Greeks. In general, however, Descartes' goal was to develop geometric methods for solving algebraic equations. He started with geometric curves and "constructed" the equations satisfied by the points on the curve.

Fermat's point of view was different. He discovered that algebraic equations actually determine geometric curves. Fermat thought of the conic sections as the curves *defined* by certain algebraic equalities. His goal was to gain important information about the geometric properties of these curves by studying the algebraic properties of the equations.

Following the pattern of our definitions of the straight line and circle we shall make the formal definitions of the conic sections in terms of the loci of the points on them. (They could also be formally defined in terms of sections of cones or in terms of algebraic equations.)

DEFINITION | A **conic section** is a locus of points whose distance from some fixed point F is a positive constant e times the distance from a fixed line l.

The fixed point is called the **focus,** the number e is called the **eccentricity,** and the fixed line is called the **directrix.** The following discussion of the different types of conic sections should help you to understand the definition better.

Parabolas

The conic section that is encountered most often in mathematics is the parabola. The path of a projectile is a parabola (unless it is fired with a force sufficient to send it into an elliptical orbit).

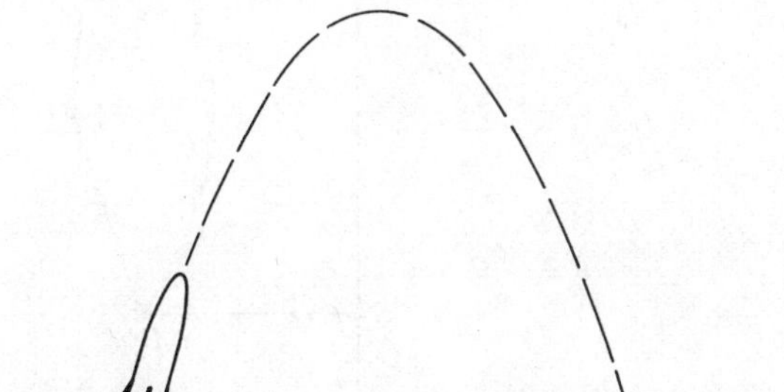

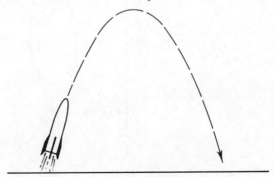

The reflectors for flashlights and the headlights of vehicles have a parabolic cross section. They are made this way because the rays from a light source placed at the focus will be reflected in parallel lines. This maximizes the amount of light projected.

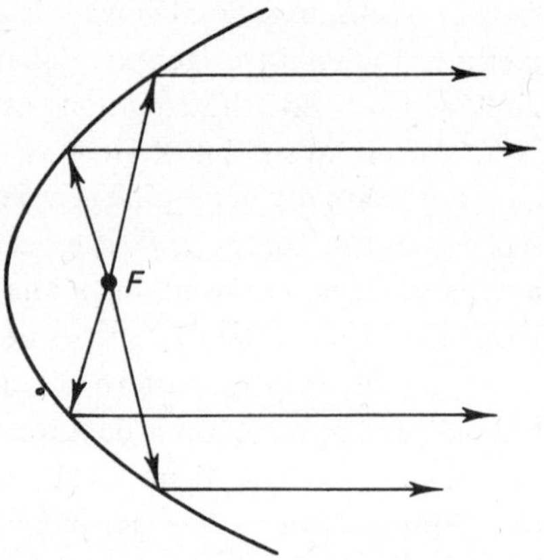

Similarly, the large antennas for the reception of radio waves and the light-gathering mirrors for astronomical telescopes are parabolic surfaces. The rays coming from distant objects are almost parallel and will be concentrated at the focus by reflection.

DEFINITION | A conic section that has eccentricity equal to 1 is called a **parabola.**

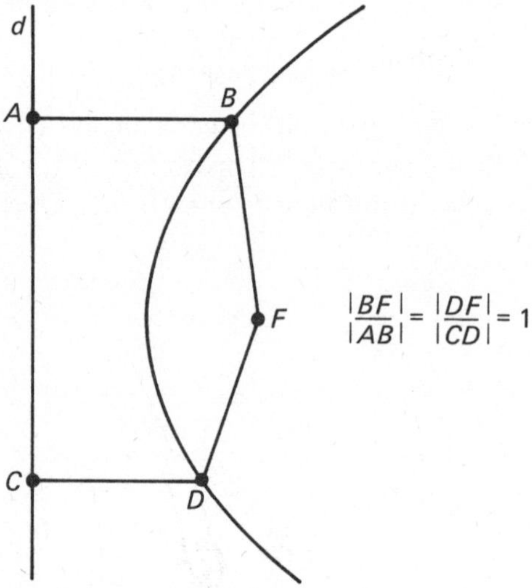

The methods of analytic geometry can be used to find algebraic conditions that must be satisfied by the points on a parabola. The condition in this case is an equation that says that the distance from the focus is equal to the distance from the directrix.

Example: By definition the set of points that are equidistant from the point $(-3, -2)$ and the line $y = 4$ form a parabola.

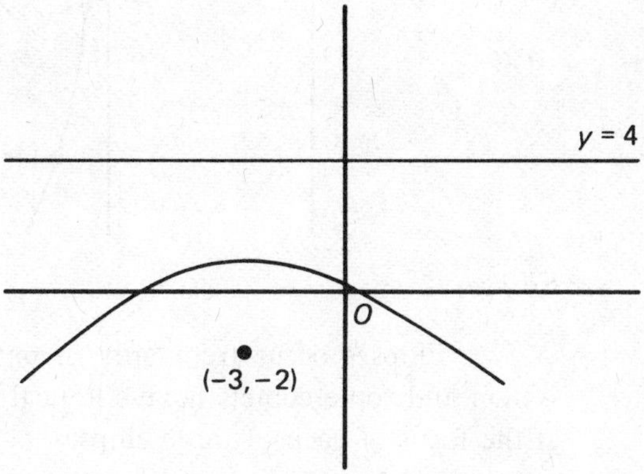

If the point (x, y) is to lie on this parabola, then the following equations must be satisfied.

$$\sqrt{(x+3)^2 + (y+2)^2} = 4 - y$$
$$(x+3)^2 + (y+2)^2 = (y-4)^2$$

The second form of the equation can be simplified further to obtain

$$(x+3)^2 = -12(y-1).$$

Example: The parabola with focus at $(5, -2)$ and directrix the line $x = -4$ has the equations

$$(x-5)^2 + (y+2)^2 = (x+4)^2$$
$$(y+2)^2 = 18(x - \tfrac{1}{2})$$

The points on the graph below were found and can be verified by choosing convenient values of y and calculating the corresponding value of x by substitution in the equation.

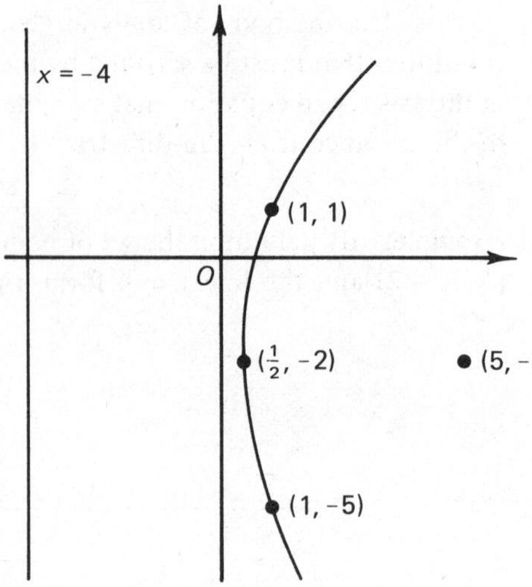

Ellipses

Ellipses occur frequently in nature. All the planets in our sol system and some comets have elliptical orbits around the sun. The sun at the focus of each of these ellipses.

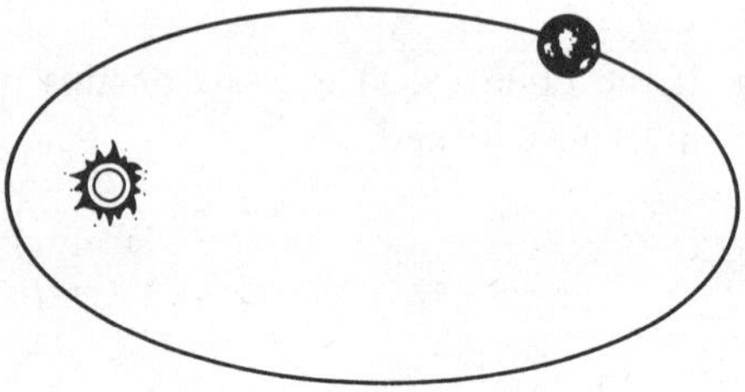

The ellipse, like the parabola, has a reflecting property that mak it useful in practical applications. Rays from the focus will be reflected by t ellipse in such a way that they converge at a point (actually another focu that is symmetrically located within the ellipse. It is this property th makes possible the "whispering galleries" in which slight sounds made at o of two points in an elliptical room can be heard clearly at the other poi

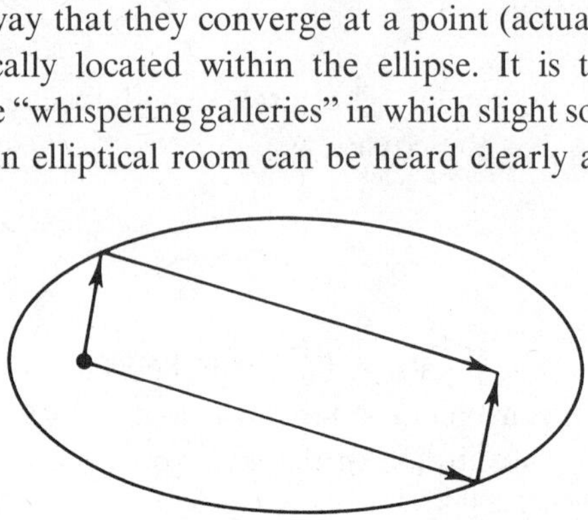

Johann Kepler. *Kepler was mathematician and astronomer for the Holy Roman Emperor Rudolf II. He founded the modern study of astronomy. (George Arents Research Library)*

DEFINITION | A conic section that has eccentricity less than 1 is called an **ellipse.**

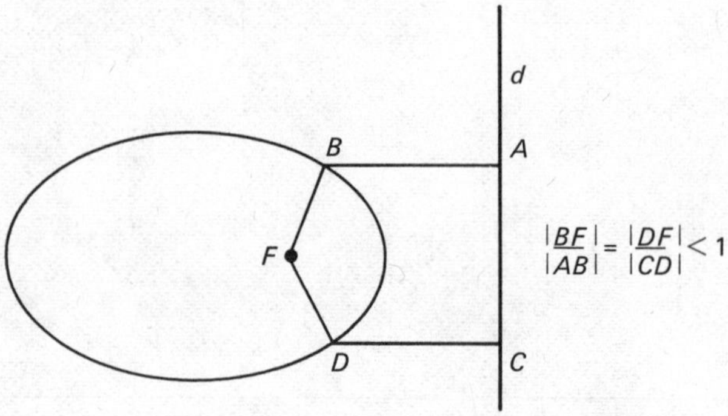

Example: The set of points that are half as far from the point (2, 0) they are from the line $x = 8$ is an ellipse. These points must satisfy t equations

$$\sqrt{(x-2)^2 + y^2} = \tfrac{1}{2}(8 - x)$$
$$(x-2)^2 + y^2 = \tfrac{1}{4}(x-8)^2$$

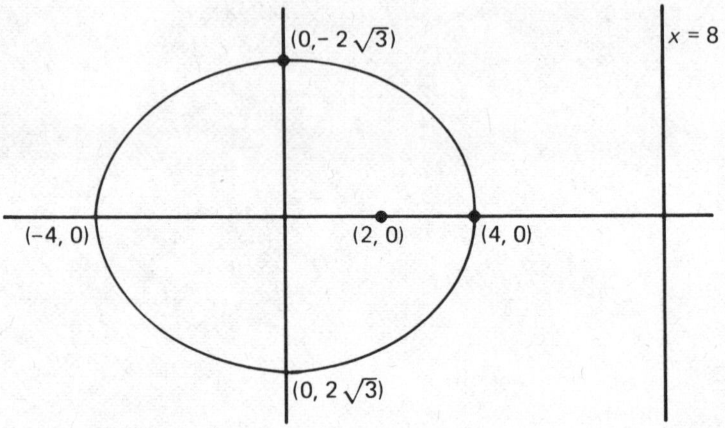

This equation can be simplified to the form $3x^2 + 4y^2 = 48$ and points the ellipse can be found by substitution.

Example: The ellipse with focus at $(-2, 1)$, directrix $y = 6$, and eccentric $\frac{1}{4}$ has the equations

$$\sqrt{(x+2)^2 + (y-1)^2} = \tfrac{1}{4}(6 - y)$$
$$16(x+2)^2 + 16(y-1)^2 = (y-6)^2$$

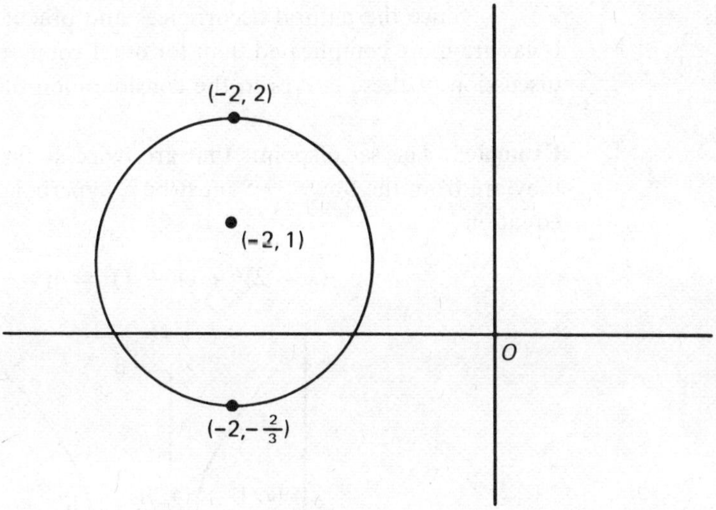

Hyperbolas

The ellipse and parabola were defined to be the conic sections with eccentricity less than and equal to the number 1. All other conic sections are hyperbolas.

DEFINITION | A conic section that has eccentricity greater than 1 is called a **hyperbola.**

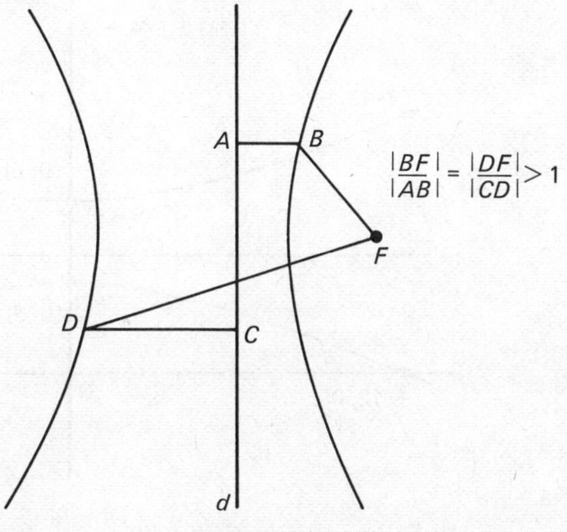

Since the natural occurrences and practical applications of hype bolas are more complicated than for other conic sections, we shall limit o discussion of these curves to the consideration of a few examples.

Example: The set of points that are twice as far from the point (2, 1) they are from the line $x = 5$ must be a hyperbola. These points satisfy th equation

$$(x - 2)^2 + (y - 1)^2 = 4(x - 5)^2$$

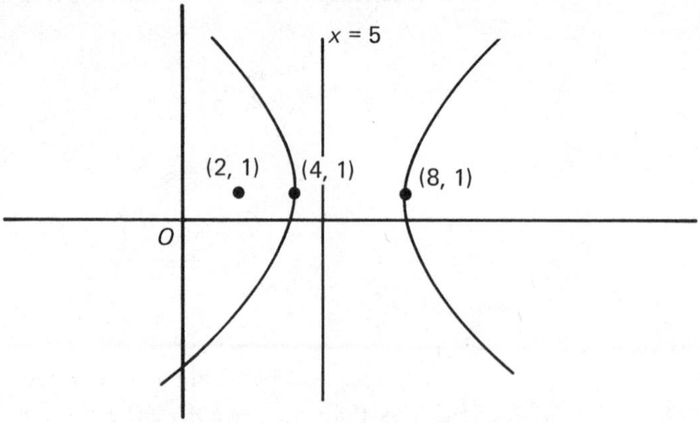

Example: The hyperbola with focus at (0, 0), directrix $y = 4$, and eccen tricity 3 has the equation

$$x^2 + y^2 = 9(y - 4)^2$$

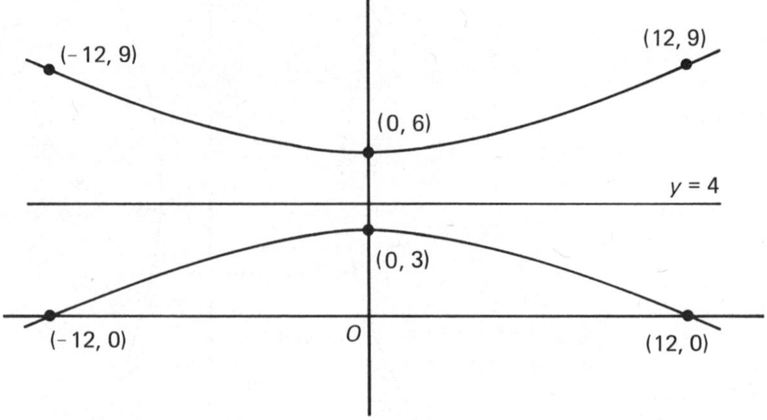

Summary

The simplest geometric curves other than straight lines are the conic sections. These curves can be defined in terms of plane sections of a cone, algebraic equations, or loci of points depending on certain distances. As a locus a conic section is a set of points for which the ratio of the distance from a fixed point to the distance from a fixed line is constant. The constant is called the *eccentricity* of the conic section. A conic section is called an *ellipse,* a *parabola,* or a *hyperbola* depending on whether its eccentricity is less than, equal to, or greater than 1.

EXERCISES

1. Identify by name the locus of points satisfying each of the following equations.

$$(x-4)^2+(y+1)^2=(y-3)^2$$
$$2x+3y=6$$
$$(x-2)^2+(y-3)^2=16$$
$$(x-4)^2+(y+1)^2=\tfrac{1}{4}(y-3)^2$$
$$(x-4)^2+(y+1)^2=9(y-3)^2$$
$$(x-4)^2+(y+1)^2=9$$
$$(x+3)^2+(y-5)^2=(x-7)^2$$
$$y=3x+7$$
$$4(x-5)^2+4(y-2)^2=(y+1)^2$$
$$16(x-1)^2+16(y-1)^2=25(y-6)^2$$

2. Check your answers to Exercise 1 by making rough sketches of the graphs of the equations.

3. If the x-axis represents the ground and a projectile fired from the origin follows the path $y = 8x - x^2$, where will the projectile land, and what is the maximum height attained?

DISCUSSION QUESTIONS

1. Draw a line and a point on a piece of paper. Use these objects as the foci and directrices of conic sections with a variety of eccentricities. What can be said about the form of a conic section which has eccentricity that is

(a) Very small?
(b) Less than 1 but very close to 1?
(c) More than 1 but very close to 1?
(d) Very large?

2. Identify the collection of curves that are the sets of points suc that the sum of the distances from two fixed points is a constant. (Such curve can be drawn by fixing the two ends of a piece of string and using a penc to trace the curve obtained by moving the pencil so as to keep the strin taut.) What happens to the shape of the curve when the two fixed point are moved closer together?

3. Identify the collection of curves that are the sets of points suc that the difference between the distances to two fixed points is constan Experiment with a variety of placements of the points and values of th constant.

4. Consider the collection of curves that have points satisfyin equations of the type $Ax^2 + Bx^2 + Cx + Dy = E$ where A, B, C, D, an E are constants. Identify the kinds of curves obtained when

(a) Both A and B are zero.
(b) Either A is zero or B is zero but not both.
(c) A and B are equal and not zero.
(d) A and B are unequal and both positive or negative.
(e) A is positive and B is negative or vice versa.

:en

he modern view of geometries

We have now considered enough examples to attempt to answer the question "What is a geometry?".

One famous mathematician of this century said, in part, "To construct a geometry is to state a system of axioms and deduce all possible consequences from them. All systems of pure geometry . . . are constructed in just this way. Their differences . . . are differences not of principle or of method, but merely of richness of content and variety of application . . . ."[1] This is an accurate answer but is too inclusive. The modern trend is to do all mathematics axiomatically. This definition of geometry does not adequately distinguish it from the rest of mathematics.

A more specific definition for geometry was given by Felix Klein (1849–1925). Klein became professor of mathematics at the University of Göttingen in 1886. During the next forty years Göttingen became the most important center of mathematical research in the world. Many of the best American mathematicians of the twentieth century went to Europe to receive their advanced training from Klein and his colleagues.

In 1872 Klein gave an address in which he presented his ideas about the properties that distinguish geometries from other kinds of mathematics. Furthermore, he showed how the various geometries can be described and differentiated in terms of these concepts. Since Klein presented his paper at the University of Erlangen, it is now referred to as the *Erlanger Programm.*

[1] G. H. Hardy, "What is Geometry?" *Math. Gazette,* XII (1925), pp. 309–316.

Transformations

In order to explain Klein's description of geometry it is necessar to formally introduce another of the basic objects of study in moder mathematics. Klein decided that the thing that all geometries have i common is that they are the investigation of what happens to the propertie of "figures" when they are "mapped" into new "figures." Since he was relatively modern mathematician, Klein was very precise about how he sai and explained this idea. What follows retains the spirit and most of th precision of the modern concept of what constitutes a geometry.

DEFINITION | A **transformation of a set *R* onto a set *S*** is a mapping of the set R to th set S in such a way that every point in R is mapped to exactly one poin of S and every point of S is the image of exactly one point of R. If R an S are the same set, then the transformation is said to be *on* S (or on R A synonym for transformation is **one-to-one correspondence.**

We shall be particularly interested in the transformations of set of points. The perspectivities and projections that were discussed in th chapter on projective geometry are examples of this kind of transformatio They are rules that assign to each point of a line or plane exactly one poin on another line or plane.

In the course of the discussion of projections a distinction was mad between metric and descriptive properties of geometric figures. It was note that the image of a figure might not have the same metric properties tha the figure possesses. Projections do, however, preserve the descriptive prop erties of geometric figures. Before we consider further examples of trans formations, it is appropriate that we formalize the concept of a trans formation leaving a property of figures unchanged.

DEFINITION | A property is **invariant** under a transformation on the set S if any subse of S having the property is mapped by the transformation onto an imag subset that also has the property.

Let us now consider some examples of typical transformations. Eac of the following examples is a transformation of the Euclidean plane ont itself. We shall use the geometric notation introduced earlier. In additio it will be convenient to use the idea of directed line segments and angle The notation $\overline{AB}$ will denote the segment from A to B and is not the sam as $\overline{BA}$. The angle $\angle\overline{ABC}$ will be the angle with "initial" side AB and "terminal" side BC.

Example: A simple kind of transformation of the plane is that which maps every point a fixed distance in the same direction. More formally, a **translation** is a transformation such that the line segments between points and their images are all equal (same directed length) and are parallel. There is a translation that will map any given point to any other.

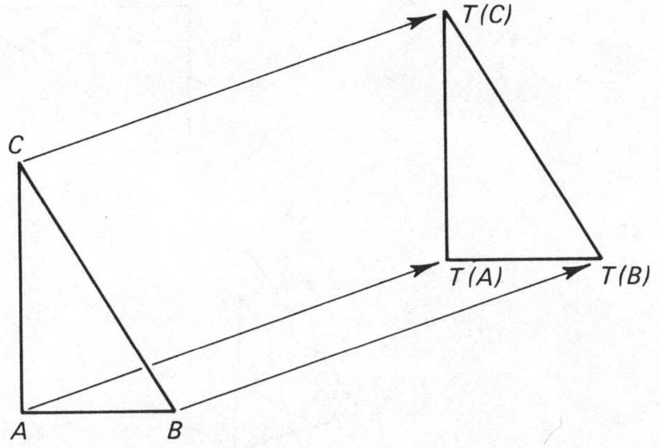

The diagram above illustrates the effect of a translation on a triangle. An arrow has been drawn from each vertex to its image. Notice that the distance between the images of two points will always be the same as the distance between the points. Transformations that have this property are of particular importance and have been given a special name.

DEFINITION | A transformation of the Euclidean plane is called an **isometry** if it leaves the distance between any two points invariant.

Example: Some other transformations are called reflections. Given a line in the plane, each point is mapped into its "mirror image" in the line. More precisely, a **reflection in the line *l*** is a transformation such that l is the perpendicular bisector of the line segment between every point and its image. All reflections in lines are isometries.

Example: A **rotation about the point *P*** is a transformation such that if $T(A)$ is the image of point A, then the lengths $|PA|$ and $|PT(A)|$ are equal and the directed angle $\angle\overline{A \cdot P \cdot T(A)}$ is the same for all points A. A rotation is defined by giving the center P of the rotation and the angle of rotation. All rotations are isometries.

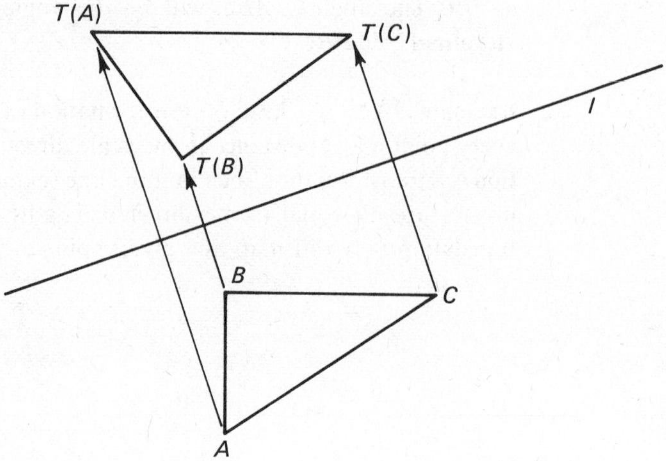

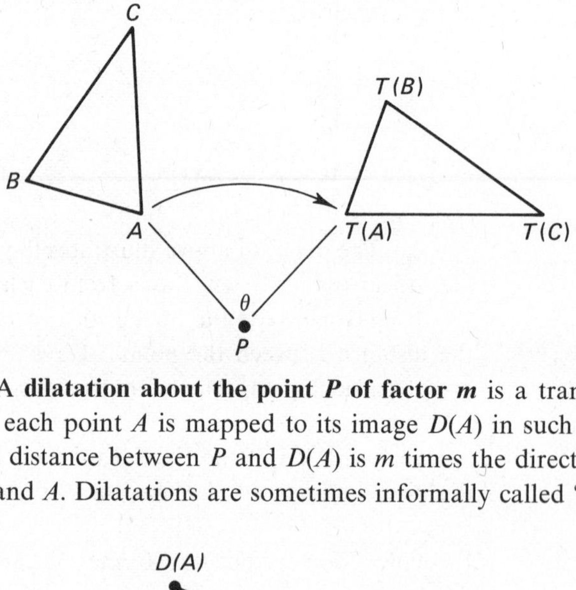

Example: **A dilatation about the point *P* of factor *m*** is a transformati D in which each point A is mapped to its image $D(A)$ in such a way th the directed distance between P and $D(A)$ is m times the directed distan between P and A. Dilatations are sometimes informally called "magnific tions."

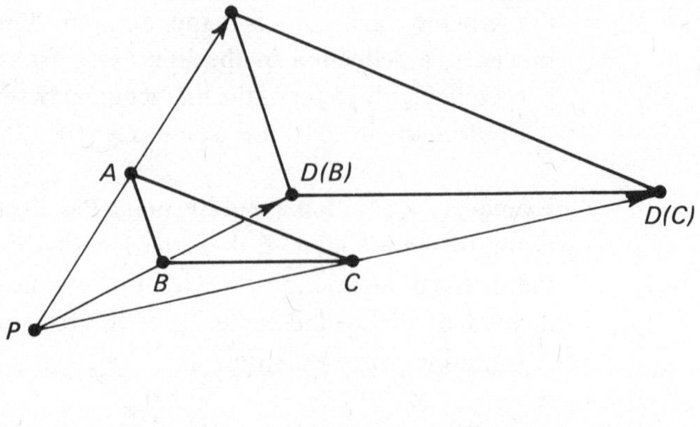

The diagram above shows a dilatation of factor 2. If the direction of the arrows were reversed, it would show a dilatation of factor $\frac{1}{2}$.

A dilatation is not an isometry unless $m = 1$ (in which case every point is mapped to itself).

Transformations are actually a very special kind of mapping. Since a transformation is one-to-one, there is another transformation that will "undo" what the transformation has "done."

DEFINITION | If T is a transformation from the set S onto the set R and $T(A)$ denotes the image of A, then the transformation from R onto S that maps $T(A)$ onto A is called the **inverse** of T.

The inverse of T is usually denoted T^{-1}. The definition of T^{-1} says that if $T(A) = B$, then $T^{-1}(B) = A$. More concisely, T^{-1} is the transformation such that $T^{-1}(T(A)) = A$.

It should be clear that diagrams for the inverses of each of the transformations illustrated above can be obtained by simply reversing the direction of the arrows. The inverse of a translation is another translation. The inverse of a rotation is a rotation with the same center but with the angle of rotation in the other direction. The inverse of a dilatation of factor m is a dilatation of factor $1/m$. The inverse of any reflection is the same reflection.

The Erlanger Programm

Klein stated that the axiomatic systems that can be properly called geometries should be selected not on the basis of the nature of the axioms but on the content of the theorems proved. He believed that the most important characteristic of geometries is that they are investigations of the effects of transformations.

DEFINITION | A **geometry** is the study of the properties that are invariant when the subsets of a set S are mapped by the transformations of some group of transformations.

(The word *group* is used here as a synonym for *set*. In Chapter 24 a special meaning is assigned to the word *group* and the definition of geometry is more specialized with that interpretation.)

Euclidean geometry is generally thought of as the study of propertie associated with the concepts of congruence and similarity. These includ such things as the property of a figure being a square, a circle, an isoscele triangle, or a parallelogram. Since an isometry will map a figure to an imag that is congruent, the set of transformations studied includes isometrie

A transformation need not, however, be an isometry for the imag of every figure to be similar to the original. The transformation need onl preserve the size of angles and the ratio of all pairs of lengths. Since th transformation must do this for all figures, it is necessary, as well as sufficien that it change each length by the same (multiplicative) factor.

DEFINITION | A transformation on the Euclidean plane is called a **similarity** if there i a positive number k such that the distance between the images of any tw points is k times the distance between the points.

By definition a dilatation with center P of factor m maps every poir to an image that is m times as far from P. The diagram below can be use to help prove that every length is multiplied by m and that a dilatatio is therefore a similarity.

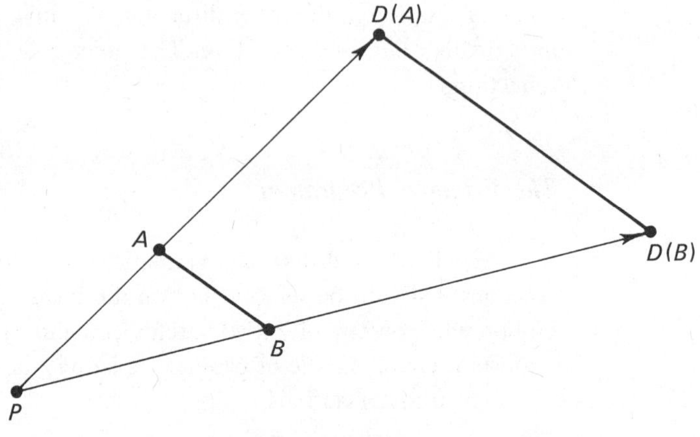

If D is a dilatation with center P of factor m, then $|P \cdot D(A)| = m|P \cdot A|$ an $|P \cdot D(B)| = m|PB|$. The triangles PAB and $P \cdot D(A) \cdot D(B)$ are then simila because of the common angle at P. It follows that $|D(A) \cdot D(B)| = m|AB$

More complicated examples of similarities and isometries can b constructed by combining two or more of the examples of this chapter. Fo example, when two similarities are done successively, the net result is als a similarity. The theorems of modern Euclidean geometry show, howeve the interest of mathematicians in reducing things to their simplest form. The most important theorems are the ones that prove that complicated transformations can actually be accomplished by doing a combination of simple transformations. The following result is typical and particularly elegant.

THEOREM | Every isometry is a combination of at most three reflections in lines.

Summary

Felix Klein originated the modern procedure for defining geometry and describing the differences between geometries in terms of transformations and the properties left invariant by the transformations. From this point of view Euclidean geometry is seen to be the investigation of properties invariant under isometries and similarities. Projective geometry and each of the non-Euclidean geometries is also the study of the effects of a special type of transformation.

To conclude our discussion on geometries, let us consider what we can infer from our knowledge of geometry about the existence and nature of other branches of mathematics. This will serve partly as a summary of things we have learned and also as guide and motivation for some of the topics to be considered in the subsequent chapters.

We have stated several times that mathematics is today done axiomatically. Euclid's development of geometry had many logical defects and his axioms and definitions have been replaced by more rigorous ones developed at the end of the nineteenth century. We have seen that to do geometry formally one must list undefined terms, state axioms and definitions, and then deduce theorems from them. This is true of all formal mathematics.

One point should probably be emphasized here. When mathematics is presented as a system with axioms and the theorems are proved with elegant logical purity, the reader is seeing the complete, polished, final product. Lost to the sight of the student, and too seldom mentioned, are the practical problems or intellectual curiosity that motivated its development, the pictures that were drawn to guide creative insight, the computations that were made that led to the formulation of conjectures, the conjectures that were tried and rejected, the inspired ideas behind the correct proofs, the years or centuries of refinement that have culminated in the particular choice of axioms and method of proof, and the applications that make the completed mathematical system useful and important.

The student who studies mathematics by memorizing the definitions and proofs of theorems is like a music student trying to understand a symphony by memorizing it note by note. The mathematician who creates the work is like the composer; he writes down every note carefully and precisely but when he thinks about his work he thinks about the sound and effect of the music being played. It should be realized by the student that the mathematician may write mathematics with great precision and logical rigor but he usually thinks about mathematics and creates it in a very informal and intuitive way. In order to understand the mathematician's appreciation of the subject and appreciate it himself, the student must look beyond the details to see how and why it was created.

It seems clear that there must be a part of mathematics devoted to the study of *logic*. The mathematician needs a good working knowledge of logic in order to do mathematics. The mathematical logician, however, is concerned with questions of independence and consistency of axioms and the kinds of theorems that are provable from given sets of axioms.

A closely related branch of mathematics is the *theory of sets*. Sets—or collections or classes—of objects are the raw materials from which mathematical systems are built. It is essential that the mathematician be able to deal with and manipulate them. Beginning with the work of George Boole (1815–1864), Georg Cantor (1845–1918), and J. W. R. Dedekind (1831–1916) and continuing well into the twentieth century, the theory of working with sets has been extensively studied. Most of mathematics was reconstructed using the theory of sets as a starting point. This is one of the reasons that there is now a "modern math" or "new math" being taught in the schools that emphasizes sets as the basic objects in mathematics.

The development of other fields of mathematics was influenced by the discovery of more subtle flaws in Euclid's logic. For example, Euclid assumed that if a line is drawn joining a point inside a circle with a point outside, then the line will intersect the circle. This seems obvious, but consider the simple case where the circle is $x^2 + y^2 = 1$ and the line is $y = x$. If they intersect, then it occurs when $2x^2 = 1$. This requires a number that might be expressed $x = 1/\sqrt{2}$. It is not clear that such a quantity is an acceptable number. Moreover, we can see now that it will be necessary, in order to avoid such difficulties, that every point on each coordinate axis be associated with some number. This realization led to the expansion of the set of numbers used by mathematicians and caused the investigation of the properties of these new *real numbers*. This is a small part of the *theory of numbers*.

We have learned that geometry is the study of invariant properties of sets of transformations. This led to the study of special kinds of sets of transformations and their relationship to special subsets of the set being mapped. These investigations are now properly part of branches of mathematics called *abstract algebra* and *linear algebra.*

Finally, the particular kinds of transformations considered in the geometries we have looked at are comparatively simple. It is a natural question to ask about the geometries obtained by working with more general transformations. *Topology* is a branch of mathematics devoted to the investigation of very general mappings and has been the subject of much mathematical research in the last half century. There are many interesting elementary curiosities in topology. Excellent nontechnical explanations of them can be found in articles such as the one by Tucker and Bailey in the book edited by Kline.[2]

EXERCISES

1. Given two points, is it always possible to map one to the other using just

(a) One translation?
(b) One reflection?
(c) One rotation?

2. Under what conditions is it possible to map a line segment onto another line segment of the same length by using only

(a) One translation?
(b) One translation and one rotation?
(c) Two reflections?

3. Show that a triangle can be mapped to any congruent triangle by using at most three reflections. (*Hint.* Use the solutions of the preceding exercises as a guide.)

***4.** If the plane is coordinated as in Chapter 8, then a transformation T can be defined by giving the coordinates of the image of every point. For example, T_1 such that $T_1(x, y) = (x + 2, y)$ or $T_1:(x, y) \rightarrow (x + 2, y)$ is a translation a distance of 2 to the right and T_2 defined $T_2(x, y) = (x, 3y)$ is a "vertical stretch" by a factor of 3. Describe the effect of each of the following transformations.

[2] Morris Kline, ed., *Mathematics in the Modern World* (New York: W. H. Freeman and Co., 1968).

T(x, y)	*Similarity?*	*Isometry?*	*Description*
(x, y)	yes	yes	maps everything onto itself
$(x, -y)$	yes	yes	reflects about the x-axis
$(-x, y)$			
$(-x, -y)$			
$(x + 2, y)$			
$(x, y - 3)$			
$(2x, 2y)$			
$(2x, -y)$			
$(x - y, x + y)$			

DISCUSSION QUESTIONS

1. Explain why the net result of doing two similarities is a similarity. Is the inverse of a similarity also a similarity? Are these statements true for isometries?

***2.** We have seen that a triangle can be mapped to any other congruent triangle by using at most three reflections. Are there other small sets of simple isometries that can be combined to accomplish this?

part three

arithmetic

eleven

the fundamental theorem of arithmetic

The kind of numbers first used by man were, of course, the counting numbers 1, 2, 3, . . . , which are today called the positive integers or the natural numbers. Records that have survived from the early civilizations show that the techniques were known for combining numbers by addition and multiplication to form new numbers. The Egyptians also accepted the reciprocals $\frac{1}{2}$, $\frac{1}{3}$, $\frac{1}{4}$, . . . of the natural numbers as "true" numbers and used them in their procedure for division. They probably viewed division of a number by 2, for example, as multiplication by $\frac{1}{2}$. The Babylonians, who do not seem to have been as skillful as the Egyptians when doing geometry, were far superior when it came to computing with numbers. One of the reasons for this is that they had a better system for denoting their numbers. The Babylonians divided any natural number by any other and had notation and techniques for working with the fractions that resulted. We shall consider later some of the methods that they developed. The idea of "negative" numbers was not used by the Egyptians, the Babylonians, or the Greeks who followed later.

Historical evidence suggests that there has been a certain amount of mysticism and superstition associated with the natural numbers. Even today it is common to think of some numbers as lucky and others as unlucky. Most of the legends and traditions that attribute mystical properties to numbers can be traced back to Pythagoras and his followers. The Pythagoreans believed the natural numbers to be the only true numbers. They regarded the natural numbers as sacred and their motto is supposed to have been "All is Number."

The Pythagoreans did not accept fractions as numbers but they were very much interested in the ratios of natural numbers. The figure on the left below is called a *star pentagon* and was the symbol of the Pythagoreans. They regarded it as special because of the aesthetically pleasing way the

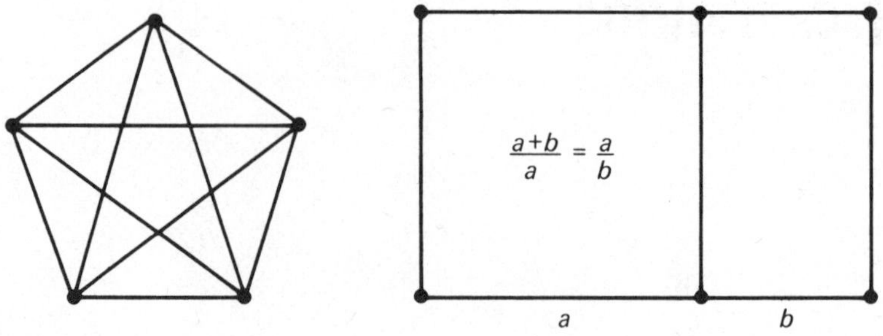

line segments are divided. Each of the diagonal lines is divided by the others into two pieces that are in what was later called the "divine proportion." In each case the ratio of the longer piece to the shorter is the same as the ratio of the whole diagonal to the longer piece.

The Pythagoreans knew about the connection between proportions and musical harmony, and they studied both subjects. In fact, they regarded music, harmony, and number as related concepts and thought a thorough knowledge of all three to be necessary for the purification of the soul.

The Greeks also treated the natural numbers logically. They multiplied and added them with ease. Subtraction was done, where possible, with no difficulty. To subtract m from n, it is necessary and sufficient to find the natural number k such that $m + k = n$. The problem of dividing one number by another is more difficult. A natural number is not exactly divisible by every other natural number. It is not possible, as it is with subtraction, to deduce from the notation for two numbers whether or not the smaller will divide the larger.

To clarify this point, let us consider the numbers 37 and 18,326. If the only numbers available are the natural numbers, then clearly 37 and 18,326 can be added and multiplied and 37 can be subtracted from 18,326. The number 18,326 cannot be subtracted from 37 since there is no natural number n such that $18{,}326 + n = 37$, and 37 cannot be divided by 18,326 since there is no m such that $(18{,}326)m = 37$. It is not obvious however, whether or not 37 divides 18,326.

A relatively sophisticated technique is necessary to decide whether the division can be carried out and to find the quotient when it can be done. In considering this problem the idea of indivisible or prime numbers arises naturally.

DEFINITION | A number, greater than 1, is said to be **prime** if its only divisors are 1 and itself.

It seems clear that every natural number greater than one can be written as a product of prime numbers. This fact provides a crude method for deciding whether one number divides another and for finding the quotient. Each number could be written as a product of its prime factors, say $m = p_1^{a_1}p_2^{a_2} \ldots p_k^{a_k}$ and $n = q_1^{b_1} \ldots q_k^{b_k}$, where the p's and q's are primes and the exponents show the number of times the prime appears as a factor. Then m divides n if every prime that divides m also divides n at least the same number of times, and the quotient is the product of the primes left over. For example, 12 divides 180 and the quotient is 15 since $12 = 2^2 \cdot 3$ and $180 = 2^2 \cdot 3^2 \cdot 5$.

Conversely, we would like to conclude that m does *not* divide n when m contains some prime as a factor more times than n does. This conclusion depends on the assumption that m and n do not have other factorizations into different combinations of primes. For example, we could be certain that 14, $2 \cdot 7$, does not divide 180 if we knew that every prime factorization of 14 must contain a 7 and no factorization of 180 ever contains a 7.

It may seem apparent that it is not possible for a natural number to have two different prime factorizations. For centuries this fact was accepted as obvious. In the nineteenth century, however, very similar situations occurred in the discussion of other kinds of numbers and these numbers were found to have more than one factorization. This discovery contributed to the desire for a more careful logical discussion of numbers and arithmetic. As we have previously noted in our discussion of geometries, such a formalization of the foundations of mathematics was done in the late nineteenth and early twentieth centuries.

The Fundamental Theorem

In keeping with the modern spirit of mathematics, and because it will enable us to show some of the ingenious ideas of the Greeks, we shall consider briefly the proof of the following theorem.

FUNDAMENTAL THEOREM OF ARITHMETIC | Every natural number greater than one can be written as the product of a finite number of primes in exactly one way.

The statement of the theorem contains two assertions. The first is that every natural number has at least one prime factorization. The second is that the prime factorization is unique. In order to prove these assertions we must make use of some of the important properties of natural numbers.

One of the basic facts about natural numbers, one which is often used as an axiom in a formal development, is the

Well-Ordering Axiom: Every nonempty set of natural numbers contains a smallest number.

This fact or its equivalent is used continually in proofs of theorems about natural numbers. The usual pattern of the proof is to show that the set of numbers for which the theorem is false is empty because it does not have a smallest element. The W.O.A. can be used to show that every natural number can be factored into primes.

Suppose that there are natural numbers that do not have prime factorizations. By the W.O.A. there must be a smallest such number. Call this number n. Since a prime number has a trivial factorization, n is not a prime. But n cannot be the product of two smaller numbers either. If $n = mk$ with $1 < m < n$ and $1 < k < n$, then m and k each have prime factorizations. A prime factorization could be constructed for n by combining those for m and k. The assumption that n is the smallest number that has no factorization leads to the conclusion that n can, in fact, be factored into primes. This is a contradiction. Therefore, every natural number has a prime factorization.

The W.O.A., or its equivalent, is also necessary to prove that there is only one factorization for each natural number. The procedure is to show that there can't be a smallest number with two different factorizations. Suppose n were such a number. If the two factorizations had any factor, say p, in common, then it could be canceled out leaving two different factorizations for the number n/p, which is smaller than n. This, of course, shows that there is no smallest number with two different factorizations. Some additional facts about natural numbers are needed here to make sure that there is a common prime in any two factorizations of the same number.

The Euclidean Algorithm

If a prime occurs in one factorization of a number, then it obviously divides the number. Repeated applications of the following theorem will prove that the prime must actually occur in *any* factorization of the number.

THEOREM VII-30 (Euclid) If a prime number p divides the product $a \cdot b$, then p divides a or p divides b.

As the notation above indicates, this theorem about prime numbers appears in Euclid's *Elements*. Euclid proved it by using a theorem that bears his name, the Euclidean Algorithm. An algorithm is a procedure for solving a problem. Euclid's Algorithm is a method for finding the greatest common divisor of two numbers.

DEFINITION The natural number d is the **greatest common divisor** of m and n if it divides both m and n and is larger than any other number that divides them both. If the greatest common divisor of two numbers is 1, then the numbers are said to be **relatively prime.**

THEOREM (Euclidean Algorithm) Each pair, m and n, of natural numbers has a greatest common divisor and the following method will produce it. Assume $n > m$. Divide m into n, finding the quotient q_1 and remainder r_1 with $0 \leqslant r_1 < m$. Divide r_1 into m, finding the quotient q_2 and remainder r_2 with $0 \leqslant r_2 < r_1$. Divide r_2 into r_1, and continue the procedure until a zero remainder is obtained. The last nonzero remainder is the greatest common divisor.

Euclid explained his algorithm by working with line segments. Given two segments of length m and n with $n > m$, one lays off the shorter on the longer as many times as possible. If there is no remainder, then m divides n and m is the g.c.d. If there is a remainder r_1, then the g.c.d. of m and n must also divide r_1. The g.c.d. of m and n must, in fact, also be the g.c.d. of m and r_1. Since r_1 is shorter than m, the problem has been "reduced." The "division process" is repeated until it comes out even. The last divisor—which is the last nonzero remainder—must be the g.c.d. of m and n.

Example: The g.c.d. of 114 and 390 is 6. Using (m, n) to denote the g.c.d. of m and n, the computations below show $(114, 390) = (114, 48) = (18, 48) = (12, 18) = (6, 12) = 6$.

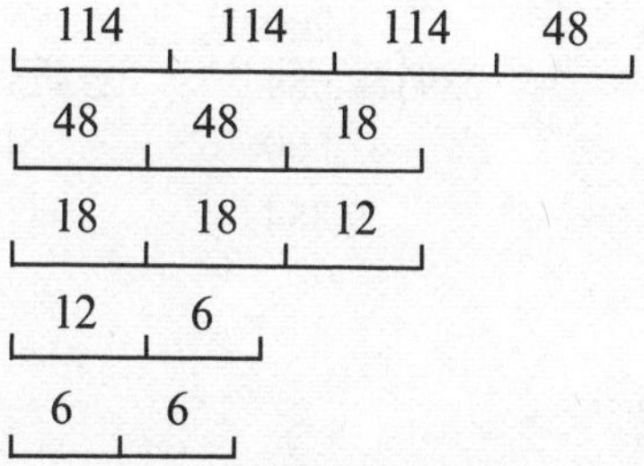

$$390 = 3 \cdot 114 + 48$$
$$114 = 2 \cdot 48 + 18$$
$$48 = 2 \cdot 18 + 12$$
$$18 = 1 \cdot 12 + 6$$
$$12 = 2 \cdot 6$$

A formal proof of the Euclidean Algorithm would require the verification of two facts. It would be necessary to prove that each division can be done and that there will in fact be a *last* nonzero remainder so that the procedure will terminate. It follows from the W.O.A. that the set of remainders does have a smallest member. The following theorem states that each of the divisions of the Euclidean Algorithm can be done.

THEOREM (Division Algorithm) If m and n are natural numbers, then it is possible to find a natural number q and a number r that is a natural number or zero such that $n = qm + r$ and $r < m$.

The Division Algorithm says that it is possible to divide n by m to find the quotient and a remainder that is smaller than the divisor. It is one more well-known property of the natural numbers that can be proved using the W.O.A.

Consider the set consisting of those numbers n, $n - m$, $n - 2m$, $n - 3m, \ldots$ that are either natural numbers or zero. This set must contain a smallest number. Let r be this smallest number and define q as the number such that $r = n - qm$. Clearly then, $n = qm + r$ and $r \geqslant 0$. If r were greater than m, then r could not be the smallest element of the set because $r - m = n - (q + 1)m$ would be in the set.

Example: Let m and n be 9 and 41. Then the numbers involved are 41, $41 - 1 \cdot 9 = 32$, $41 - 2 \cdot 9 = 23$, $41 - 3 \cdot 9 = 14$, $41 - 4 \cdot 9 = 5$, $41 - 5 \cdot 9 = -4, \ldots$. The smallest natural number here is 5 and is the remainder. Since $5 = 41 - 4 \cdot 9$, the value of q is 4.

The proof of the Division Algorithm confirms that it is always *possible* to determine whether one natural number exactly divides another. The algorithm, which consists of repeatedly subtracting the divisor, is not very practical. The commonly used procedure of "long division" is a streamlined version of this algorithm.

Example: Let m and n be 239 and 57,684, respectively. The usual computation of long division is shorthand for the following line of reasoning. Clearly 57,684 is less than $239 \cdot 1000$ but larger than $239 \cdot 100$. How many hundreds of 239's can be subtracted?

$$\begin{array}{r} 200 \\ 239\overline{)57{,}684} \\ 47{,}800 \\ \hline 9{,}884 \end{array} \qquad \begin{array}{r} 240 \\ 239\overline{)57{,}684} \\ 47{,}800 \\ \hline 9{,}884 \\ 9{,}560 \\ \hline 324 \end{array}$$

After two hundred 239's have been subtracted, the remainder is 9884, which is less than $239 \cdot 100$ but more than $239 \cdot 10$. The next step is to remove as many 10's of 239 as possible. Finally as many more 239's as possible, certainly less than 10, are subtracted to complete the computation. The quotient is 241 and the remainder is 85.

It is worth noting that all the theorems of this chapter can be proved no matter what system of symbols is used to denote the natural numbers. That is, the facts about divisibility and prime numbers are true whether the numbers are represented by Egyptian hieroglyphic characters, Roman numerals, or quantities that a computer can work with. Modern computational techniques such as long division are effective largely because of the excellence of the so-called "Arabic" system of numerals. The symbols 1, 2, 3, . . . , 9 are actually descendants of those used by the Hindus as early as the sixth century A.D. and perhaps even earlier. Later the symbol 0 was introduced as a *placeholder* and the modern positional notation was developed. Europeans learned of the notation from the Arabs and gradually adopted it about the time of the Crusades.

In the next chapter we shall see a further example of how the choice of appropriate notation can simplify computations and actually lead to the discovery of new theorems.

Summary

An important fact about the natural numbers is that each has a unique factorization into prime numbers. The Fundamental Theorem of Arithmetic can be proved by using the Well-Ordering Axiom, the Division Algorithm, the Euclidean Algorithm, and another theorem of the *Elements* that provides the basic fact about prime numbers.

Although the choice of notation for the natural numbers affects the efficiency of computational techniques, it has no bearing on the essential properties of the numbers.

EXERCISES

1. Use the Euclidean Algorithm to verify the following statements about greatest common divisors.

(a) $(38, 162) = 2$ because $(38, 162) = (38, 10) = (8, 10) = (2, 8) = 2$.

(b) $(12, 25) = (12, 1) = 1$.

(c) $(108, 831) = (108, 75) = (33, 75) = (33, 9) = (6, 9) = (6, 3) = 3$.

2. Find the prime factorizations of 812, 71, 261, 1001, and 919.

3. Is the Well-Ordering Axiom true for the set of all
(a) Positive and negative whole numbers?
(b) Fractions?
(c) Positive fractions?

DISCUSSION QUESTIONS

1. It is desirable to have a method for finding all prime numbers less than a given number. Investigate why the following example of the method called the *Sieve of Eratosthenes* works. "To find all prime numbers less than 100, write down the numbers 1–100. Take out 1 and all multiples of 2 except 2. Then take out all multiples of 3 except 3. Repeat this for 5 and 7. Everything remaining is a prime." How should the method be extended to find all primes less than 1000 or less than 10^6?

2. How many prime numbers are there?

3. Do any numbers other than primes have the property that if the number divides the product mn, then it divides either m or n?

4. Suppose that the number m has the property that if m divides the square of a number, say n^2, then m divides n. What can be said about the prime factorization of m?

***5.** The sequences of "reductions" obtained using the Euclidean Algorithm are of various lengths. Two pairs of numbers of about the same size can have sequences of very different lengths. For example, $(120, 840) = 120$ and $(112, 831) = (112, 47) = (18, 47) = (11, 18) = (11, 7) = (4, 7) = (4, 3) = (1, 3) = 1$. Find a way to construct pairs of numbers that require a maximum number of steps for their size.

twelve

divisibility and congruence

It was noted in the preceding chapter that the problem of deciding when one whole number divides another requires a great deal more effort to solve than is needed to solve the analogous problem for addition, multiplication, or subtraction. The difficulty would of course disappear if we allowed fractions as answers. Because there are many applications in which a fractional answer makes no sense, however, and more importantly (and merely) because the problem exists, we shall investigate methods for determining whether or not a given natural number divides another.

C. F. Gauss developed notation and terminology in his book *Disquisitiones Arithmeticae* (1801) that facilitate the solution of this and many other kinds of problems. We have already noted that Gauss was the first to realize that geometries other than Euclid's are possible. He reached this realization after a careful study of the logical foundations of geometry. The *Disquisitiones Arithmeticae* is the result of his investigation of the basic laws for working with numbers. In this book Gauss provided a proof of the Fundamental Theorem of Arithmetic. The book was finished while Gauss was still a student and he dedicated it to his patron, the Duke of Brunswick, who had provided the money so that he could attend school.[1]

[1] C. F. Gauss, *Disquisitiones Arithmeticae,* trans. by A. A. Clarke (New Haven: Yale University Press, 1966).

Congruence

The most important of Gauss's arithmetical achievements was his introduction of the concept of *congruence* and his invention of the notation that makes it such a powerful technique. Before stating Gauss' definition and proving his very general theorems, it is probably useful to consider a specific simple problem.

If we choose some number, say 23,456, then it is obvious that 1 is a divisor because 1 divides every natural number. It is also clear that 2 divides 23,456. The reasons for the "obviousness" of this fact are more complicated. What we have learned from experience is that 2 divides a number if it divides the last digit. We "know" this and would, if pressed, eventually explain it in detail by quoting some other "obvious" facts.

1. The notation 23,456 stands for $2 \cdot 10^4 + 3 \cdot 10^3 + 4 \cdot 10^2 + 5 \cdot 10 + 6$.

2. The number 10 is divisible by 2.

3. If 2 divides the number m, and n is any other number, then 2 divides mn.

4. If 2 divides both m and n, then 2 divides $(m + n)$.

5. If 2 divides m and does not divide n, then it does not divide $(m + n)$.

Putting these facts together we can justify our rule for divisibility by the number 2. The number 23,456 is the sum of terms that are products. Each of the terms that contains a factor of 10 is divisible by 2. The sum of these terms is divisible by 2. The number is the sum of these terms and the last digit. It is divisible by 2 if and only if the last digit is.

Gauss isolated the essential idea in this kind of argument. He then defined the concept and introduced the notation that made it possible to discover and prove very general facts about divisibility by any natural number.

DEFINITION | Let m be a natural number. The integers n and k are said to be **congruent modulo** $\boldsymbol{m}$ if m divides their difference. This will be denoted $n \equiv k \pmod{m}$.

Examples: Since $17 - 5 = 12$ and 12 is divisible by 3, 4, 6, and 12, we have

$$17 \equiv 5 \pmod{3} \qquad 17 \equiv 5 \pmod{6}$$
$$17 \equiv 5 \pmod{4} \qquad 17 \equiv 5 \pmod{12}.$$

Carl Friedrich Gauss. *Gauss is generally considered to be the greatest mathematician of all time.* (*George Arents Research Library*)

It is also easy to verify by subtraction and division that

$$7 \equiv 32 (\text{mod } 5) \qquad 8 \equiv -6 (\text{mod } 7) \qquad 31 \equiv 184 (\text{mod } 51).$$

The following special cases should be clear:

$$8 \equiv 0 (\text{mod } 4) \qquad 15 \equiv 0 (\text{mod } 3)$$

One important consequence of the definition, which is in fact an equivalent way of defining congruence, is that $n \equiv k (\text{mod } m)$ if and only if n and k leave the same remainder when divided by m. In particular if $n \equiv k (\text{mod } m)$ and m divides n, then m divides k. The number n satisfies $n \equiv 0 (\text{mod } m)$ if and only if m divides n.

The facts we have been using about divisibility by the number 2 are special cases of the rules for any divisor that are stated in the following theorem.

THEOREM | Let m be a natural number and let a, b, c and d be integers such that $a \equiv b (\text{mod } m)$ and $c \equiv d (\text{mod } m)$. Then

1) $a + c \equiv b + d (\text{mod } m)$
2) $ac \equiv bd (\text{mod } m)$.

The hypotheses of the theorem are that $a - b$ and $c - d$ are both divisible by m. The conclusions are that $(a + c) - (b + d)$ and $ac - bd$ are divisible by m. If $a - b = km$ and $c - d = pm$, then $(a + c) - (b + d) = (k + p)m$ and $ac - bd = (ac - ad) + (ad - bd) = (ap + dk)m$. The theorem is proved.

It is not easy to understand and believe these statements unless one has a clear understanding of the meaning of the notation. We shall examine the conclusions for some specific values of m.

Example: Since $7 \equiv 1 (\text{mod } 6)$, $11 \equiv 5 (\text{mod } 6)$, and $8 \equiv 2 (\text{mod } 6)$, the theorem proves that

$$7 + 11 \equiv 1 + 5 \equiv 0 (\text{mod } 6) \qquad 77 \equiv 1 \cdot 5 \equiv 5 (\text{mod } 6)$$
$$7 \cdot 11 \cdot 8 \equiv 1 \cdot 5 \cdot 2 \equiv 10 \equiv 4 (\text{mod } 6)$$
$$7^2(11 + 8^4) \equiv 1 \cdot (5 + 16) \equiv 21 \equiv 3 (\text{mod } 6).$$

More generally, if $n \equiv 2 (\text{mod } 6)$ and $m \equiv 4 (\text{mod } 6)$, then $n + m$ is divisible by 6 because

$$n + m = 2 + 4 \equiv 6 \equiv 0 (\text{mod } 6).$$

Example: Every natural number is congruent to either 0 or 1 modulo 2. If $n \equiv 0 \pmod 2$, then n is even and divisible by 2. If $n \equiv 1 \pmod 2$, then n is odd and not divisible by 2. The statements below are elementary consequences of the theorem and are also restatements of the rules for divisibility by 2 given earlier.

If $n \equiv 0 \pmod 2$ and $m \equiv 0 \pmod 2$, then $n + m \equiv 0 \pmod 2$.
If $n \equiv 0 \pmod 2$ and $m \equiv 1 \pmod 2$, then $n + m \equiv 1 \pmod 2$.
If $n \equiv 0 \pmod 2$, then $nm \equiv 0 \pmod 2$ for all m.

If the set of integers is partitioned into subsets so that two integers are in the same subset if and only if they are congruent modulo some particular number m, then the result is exactly m disjoint subsets. Each of the subsets can be identified by indicating just one of its members. It is customary to choose the smallest nonnegative number in the set. If the value of m is known, then the symbol $\underline{n}$ will denote the set of all integers that are congruent to n modulo m.

For example, when $m = 5$, the sets are

$$\underline{0} = \{0, 5, -5, 10, -10, \ldots\} \qquad \underline{3} = \{3, 8, -2, \ldots\}$$
$$\underline{1} = \{1, 6, -4, \ldots\} \qquad \underline{4} = \{4, 9, -1, \ldots\}$$
$$\underline{2} = \{2, 7, -3, \ldots\}$$

The theorem of this section shows that the sets to which the sum and product of two numbers belong depend only on the sets from which the numbers are chosen. When $m = 5$, the sum of any number from $\underline{2}$ and any number from $\underline{4}$ must belong to $\underline{1}$ and the product must be in $\underline{3}$. Since there will always be only m different sets, it is possible to summarize this kind of information in "addition" and "multiplication" tables. The tables below are for the case $m = 7$.

+	$\underline{0}$	$\underline{1}$	$\underline{2}$	$\underline{3}$	$\underline{4}$	$\underline{5}$	$\underline{6}$
$\underline{0}$	$\underline{0}$	$\underline{1}$	$\underline{2}$	$\underline{3}$	$\underline{4}$	$\underline{5}$	$\underline{6}$
$\underline{1}$	$\underline{1}$	$\underline{2}$	$\underline{3}$	$\underline{4}$	$\underline{5}$	$\underline{6}$	$\underline{0}$
$\underline{2}$	$\underline{2}$	$\underline{3}$	$\underline{4}$	$\underline{5}$	$\underline{6}$	$\underline{0}$	$\underline{1}$
$\underline{3}$	$\underline{3}$	$\underline{4}$	$\underline{5}$	$\underline{6}$	$\underline{0}$	$\underline{1}$	$\underline{2}$
$\underline{4}$	$\underline{4}$	$\underline{5}$	$\underline{6}$	$\underline{0}$	$\underline{1}$	$\underline{2}$	$\underline{3}$
$\underline{5}$	$\underline{5}$	$\underline{6}$	$\underline{0}$	$\underline{1}$	$\underline{2}$	$\underline{3}$	$\underline{4}$
$\underline{6}$	$\underline{6}$	$\underline{0}$	$\underline{1}$	$\underline{2}$	$\underline{3}$	$\underline{4}$	$\underline{5}$

·	$\underline{0}$	$\underline{1}$	$\underline{2}$	$\underline{3}$	$\underline{4}$	$\underline{5}$	$\underline{6}$
$\underline{0}$	$\underline{0}$	$\underline{0}$	$\underline{0}$	$\underline{0}$	$\underline{0}$	$\underline{0}$	$\underline{0}$
$\underline{1}$	$\underline{0}$	$\underline{1}$	$\underline{2}$	$\underline{3}$	$\underline{4}$	$\underline{5}$	$\underline{6}$
$\underline{2}$	$\underline{0}$	$\underline{2}$	$\underline{4}$	$\underline{6}$	$\underline{1}$	$\underline{3}$	$\underline{5}$
$\underline{3}$	$\underline{0}$	$\underline{3}$	$\underline{6}$	$\underline{2}$	$\underline{5}$	$\underline{1}$	$\underline{4}$
$\underline{4}$	$\underline{0}$	$\underline{4}$	$\underline{1}$	$\underline{5}$	$\underline{2}$	$\underline{6}$	$\underline{3}$
$\underline{5}$	$\underline{0}$	$\underline{5}$	$\underline{3}$	$\underline{1}$	$\underline{6}$	$\underline{4}$	$\underline{2}$
$\underline{6}$	$\underline{0}$	$\underline{6}$	$\underline{5}$	$\underline{4}$	$\underline{3}$	$\underline{2}$	$\underline{1}$

You may have noticed that the theorem can be interpreted as stating properties of the symbol $\equiv$ similar to those that we commonly associate with $=$. The first conclusion of the theorem says something analogous to "equals added to equals are equal." The theorem actually proves that certain arithmetic manipulations with congruences can be done using the familiar rules for algebraic manipulations with equations.

For example, the theorem provides a method for solving problems of the type: "Find all integers x such that 7 divides $5x - 4$." We can deduce from $5x \equiv 4 \pmod 7$ that $3 \cdot 5x \equiv 3 \cdot 4 \pmod 7$ and $x \equiv 5 \pmod 7$. The answer is, therefore, "All integers that leave a remainder of 5 when divided by 7." We do not have the space to pursue this further but the interested reader can find additional information in any book on elementary *number theory*.

Divisibility Tests and Techniques

We shall now apply the theorem of the preceding section to obtain some simple rules for divisibility. The general method will be to first decompose large numbers into combinations of smaller ones. The theorem can then be used to determine the properties of the larger one from those of the smaller ones. Since the formal proofs of these results would be tedious, we shall generalize from a few examples.

Example: Consider the problem of determining whether or not a large number is divisible by 3. Suppose, for example, that the number is 23,456. We can solve the problem by noting the following facts that are consequence of the theorem.

$$10 \equiv 1 \pmod 3 \qquad 10^2 \equiv 1 \pmod 3$$

In fact, $10^n \equiv 1 \pmod 3$ for all n.

$$2 \cdot 10^4 \equiv 2 \pmod 3 \qquad 3 \cdot 10^3 \equiv 3 \pmod 3$$
$$4 \cdot 10^2 \equiv 4 \pmod 3 \qquad 5 \cdot 10 \equiv 5 \pmod 3$$
$$23{,}456 = 2 \cdot 10^4 + 3 \cdot 10^3 + 4 \cdot 10^2 + 5 \cdot 10 + 6$$
$$23{,}456 \equiv 2 + 3 + 4 + 5 + 6 \pmod 3$$
$$23{,}456 \equiv 2 \pmod 3.$$

Thus, 23,456 leaves a remainder of 2 when divided by 3. The remainder when 478 is divided by 3 is 1 since

$$478 = 4 \cdot 10^2 + 7 \cdot 10 + 8 \equiv 4 + 7 + 8 \equiv 19 \equiv 1 \pmod 3.$$

In general, a number is congruent modulo 3 to the sum of its digits. A number is divisible by 3 if and only if the sum of its digits is divisible by 3.

Example: Consider the problem of finding the remainder when a number is divided by 9. Since $10 \equiv 1 \pmod 9$, it follows from the theorem that $10^n \equiv 1^n \equiv 1 \pmod 9$. As in the case of the number 3, we see that every number is congruent modulo 9 to the sum of its digits.

$$3472 = 3 \cdot 10^3 + 4 \cdot 10^2 + 7 \cdot 10 + 2$$
$$3472 \equiv 3 + 4 + 7 + 2 \equiv 16 \equiv 7 \pmod 9$$

Thus 3472 leaves a remainder of 7 when divided by 9.

Example: It is sometimes convenient to replace a number by a negative integer that is congruent to it. Since $10 \equiv -1 \pmod{11}$, we see that $10^n \equiv 1 \pmod{11}$ if n is even and that $10^n \equiv -1 \pmod{11}$ if n is odd. Thus,

$$3472 \equiv -3 + 4 - 7 + 2 \equiv -4 \equiv 7 \pmod{11}$$

and

$$82649 \equiv +8 - 2 + 6 - 4 + 9 \equiv 6 \pmod{11}.$$

Notice that the units digit is positive and the signs alternate as the powers of 10 increase. Every number is congruent modulo 11 to an "alternating sum" of its digits.

Even though there are no easy rules for determining divisibility by numbers such as 7, 8, and 12, similar techniques can be used to simplify the computations involved in working with large numbers.

Examples: The number 3472 is divisible by 8 since $10 \equiv 2 \pmod 8$ and

$$3472 \equiv 3 \cdot (2)^3 + 4 \cdot (2)^2 + 7 \cdot 2 + 2 \equiv 0 + 0 + 14 + 2 \equiv 0 \pmod 8.$$

The number 5623 leaves a remainder of 7 when divided by 12 since $10 \equiv -2 \pmod{12}$ and

$$5623 \equiv 5(-8) + 6 \cdot 4 + 2 \cdot (-2) + 3 \equiv -17 \equiv 7 \pmod{12}.$$

We can use the tables on p. 115 to assist with the computations that show that 5261 leaves a remainder of 4 when divided by 7.

$$5261 \equiv 5(3)^3 + 2(3)^2 + 6(3) + 1 \equiv 5 \cdot 6 + 2 \cdot 2 + 4 + 1 \equiv 4 \pmod 7$$

The theorem about congruences that we have been applying state facts that do not depend on the system of notation used for the integer It should be noted, however, that the rules we have obtained for divisibilit by numbers such as 2, 3, and 9 depend on the fact that our numbers ar written base 10 (in terms of powers of 10). They are not valid if some othe base is used.

Example: Suppose that 5261 is the base 8 representation of a natura number. Then this number is divisible by 7 and leaves a remainder of when divided by 9.

$$5261 = 5(8)^3 + 2(8)^2 + 6(8) + 1$$
$$5261 \equiv 5 + 2 + 6 + 1 \equiv 0 (\text{mod } 7)$$
$$5261 \equiv -5 + 2 - 6 + 1 \equiv 1 (\text{mod } 9)$$

Summary

Gauss developed the theory of congruences. Very general theorem about the divisibility properties of natural numbers can be proved easil by his methods. A variety of computational shortcuts are corollaries of thes theorems.

EXERCISES

1. Verify that

$$26 \equiv 42 (\text{mod } 8)$$
$$43 \equiv 19 (\text{mod } 6)$$
$$43 \equiv 15 (\text{mod } 7).$$

2. Show that

$$374 \equiv 2 (\text{mod } 3) \qquad 374 \equiv 2 (\text{mod } 4)$$
$$374 \equiv 0 (\text{mod } 2) \qquad 374 \equiv 4 (\text{mod } 5)$$
$$374 \equiv 3 (\text{mod } 7) \qquad 374 \equiv 0 (\text{mod } 11).$$

Is this information helpful in finding a prime factorization of 374?

3. State and justify a rule for whether or not a number is divisib by 5.

4. Use the results of this chapter to find the prime factorizatio of the following numbers.

(a) 111 (c) 5863
(b) 2772 (d) 623

5. Construct tables showing how remainder sets are added and multiplied modulo 3 and modulo 4.

DISCUSSION QUESTIONS

1. Determine the day of the week on which you were born. First find the day on which your birthday falls this year and then subtract appropriate multiples of 365 and 366 modulo 7.

2. We saw that the rules for divisibility depend on the fact that we write numbers base 10. Investigate to find the rules that would result if numbers are written with some other base. Is there, for example, a simple rule for divisibility by 5 or 7 when numbers are written base 6 or for divisibility by 3 when numbers are written base 2?

***3.** Suppose $n \geqslant 5$. Show by examples that the product of all numbers less than n seems to be congruent to 0 or -1 modulo n [for example, $1 \cdot 2 \cdot 3 \cdot 4 \equiv -1 (\text{mod } 5)$]. Try to find out why this happens and to discover a rule for determining which will happen for particular choices of n.

thirteen

the existence of irrational numbers

The Pythagoreans thought that the natural numbers were of divir origin and that no quantities existed in the universe that were not expressib as ratios of natural numbers. They discovered, however, that it is possib to geometrically construct lengths that are not rational multiples of the give "starting" lengths. There are a number of legends about their reactions this discovery. Most of these accounts indicate that the Pythagoreans viewe the existence of such quantities as a contradiction to their entire phil sophical view of the universe and tried to keep the result a secret. One sto is that they murdered a member of their sect named Hippasus for tellir outsiders about the existence of such "numbers."

Constructible Lengths

Lest we underestimate the early Greek arithmeticians, let us lo more closely at what they accomplished. In Euclid's *Elements* and oth works of the time, numbers were represented by line segments that we labeled with letters. Theorems about numbers were proved by showing th the conclusions held for segments or areas of the appropriate sizes. F example, the distributive law $a(b + c) = ab + ac$ and the ru $(a + b)^2 = a^2 + 2ab + b^2$ are illustrated by the diagrams that follow.

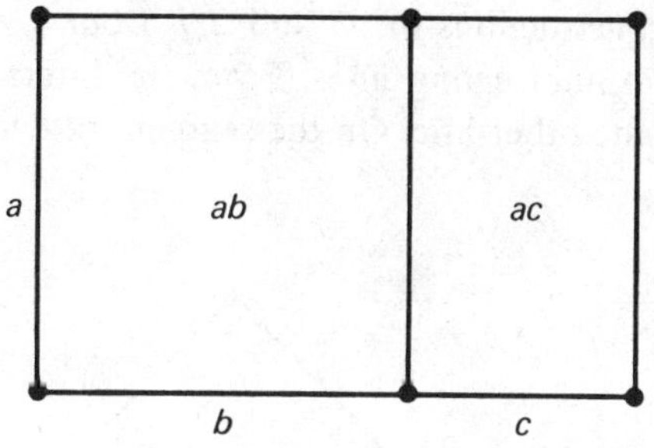

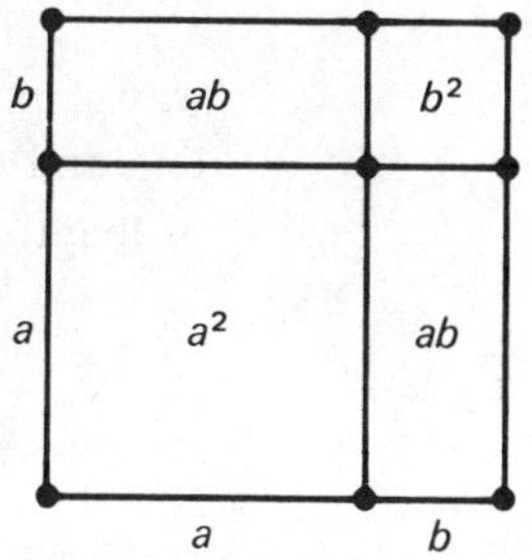

To prove these theorems it was necessary to be able to construct sums, differences, products, and quotients of the lengths given. Suppose that two segments are provided with lengths a and b that are natural numbers. Then it is easy to construct segments of lengths $a + b$ and $a - b$ (provided $a > b$).

The Greeks would have represented the product $a \cdot b$ as an area by constructing a rectangle. They rephrased division problems in terms of multiplication. In the appendix on geometry of his *Discourse,* Descartes showed how segments of length ab and a/b can be constructed if given segments of lengths a, b, and 1. A similar technique can be used to construct a segment of length 1 if given either a or b.

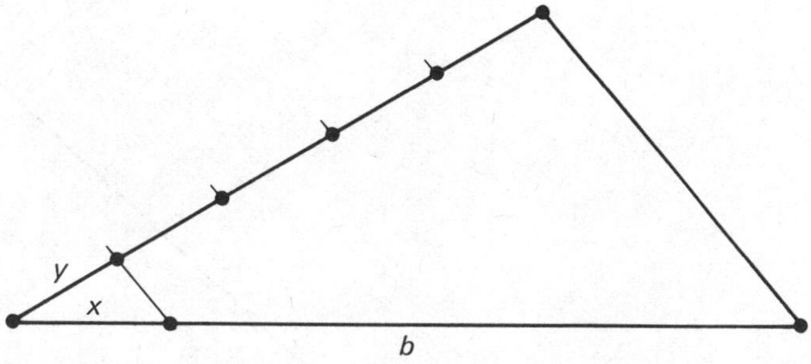

We first construct 1 from a segment of length b. Through one end of the segment of length b, draw an intersecting line. On this line, starting at the intersection, measure off b copies of some convenient length, say y. Connect the end of the original segment with the end of the segment of length yb. Construct a line parallel to this new line through the first measured length. This will cut off a segment of length x such that the resulting triangles are similar. Thus, $x/y = b/by$ and $x = 1$.

The constructions of ab and a/b from 1, a, and b start the same way. Draw two intersecting lines. From the intersection, lay off a on one line and b on the other line. On the segment of length b, lay off a segment of length 1.

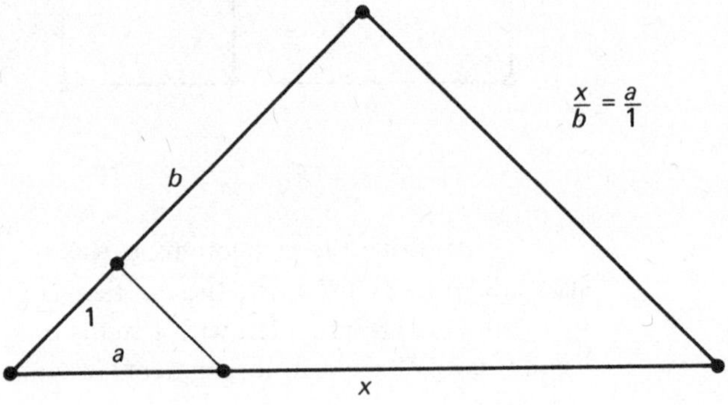

To construct a segment of length ab, connect the ends of the segments of length 1 and a; then draw a parallel to this line through the end of the segment of length b. This parallel will intersect the line containing the segment of length a at a distance of ab from the intersection of the original two lines. (The triangles are similar.)

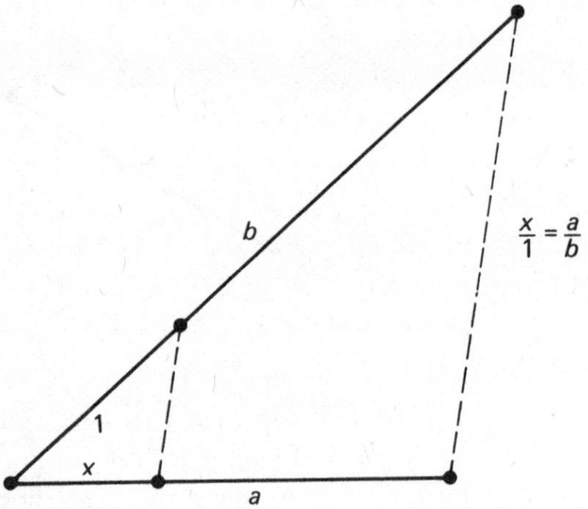

To construct a segment of length a/b, connect instead the ends of the segments of length a and b; then draw a parallel line through the end of the segment of length 1. It will intersect the other line at a/b from the intersection of the original two lines.

The first lengths that are not exactly measurable by a quotient of natural numbers probably were not deliberately constructed. The diagonal of a square of side 1, for example, is a length x such that $x^2 = 2$ by the Pythagorean Law. It is easy, however, to intentionally construct a length that is the square root of any given length. On a line, lay off 1 and n. Construct a semicircle with this segment as diameter. Draw a perpendicular at the end of the unit length. The segment of the perpendicular in the semicircle has the desired length.

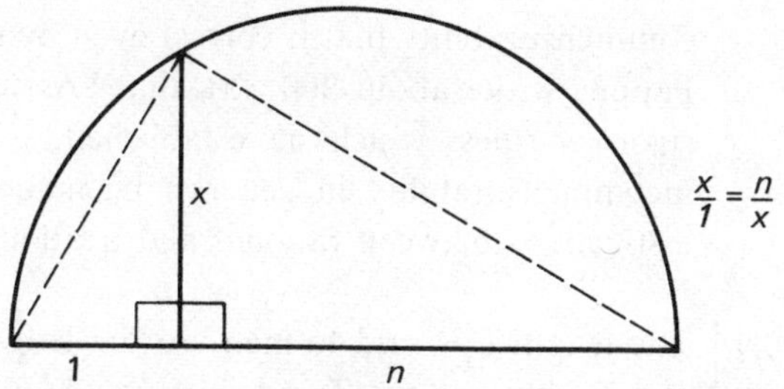

The Pythagoreans were able to *prove* that certain of these constructable lengths are not rational. The argument following the statement of the next theorem is the one they used.

THEOREM | There is no rational number whose square is 2.

Suppose that the theorem is false. Then there are natural numbers m and n such that $(m/n)^2 = 2$. We can further assume that m and n are relatively prime because it would not change their ratio to divide out common factors. We have $m^2 = 2n^2$. This certainly implies that 2 divides m^2. Since 2 is prime, we see by the basic property of primes (p. 107) that 2 must divide m. Let $m = 2k$; then we have $4k^2 = 2n^2$ or $2k^2 = n^2$. Repeating the same reasoning we see that 2 divides n. But this contradicts the fact that m and n were chosen to be relatively prime. Since our logic is correct and we reached a contradiction, the hypothesis that $(m/n)^2 = 2$ must be false. There can be no rational number whose square is 2.

The Greeks worried about the fact that a great many constructible lengths are not measurable with ratios of natural numbers. They excluded these quantities from their arithmetic because such things did not fit their definition of "number." They developed rules for working with such quantities in geometry because they were known to exist there.

Using their geometric techniques, the Greeks proved many theorems about the natural numbers and ratios of natural numbers. Most of these theorems are actually true for *any number that can be represented by a length.*

Book V of the *Elements* is devoted to a discussion of proportions, which seems to be the work of Eudoxus of Cnidus (408–355 B.C.) and which is valid whether or not the quantities are rational.

Incommensurables and Irrationals

Euclid made significant contributions to the subject of irrational numbers in the *Elements*. Book X contains an extensive discussion of incommensurability that is correct even by today's standards of logical rigor. Pappus wrote about 300 A.D. that "As for Euclid, he set himself to give rigorous rules, which he established, relative to commensurability and incommensurability in general; he made precise the definitions and the distinctions between rational and irrational magnitudes . . ."[1]

DEFINITION (Euclid) The length u is said to **measure** the length s if there is a natural number n such that $s = nu$. Two lengths s and t are said to be **commensurable** if they are both measurable by the same length and are said to be **incommensurable** otherwise. A length is **rational** if it is commensurable with 1. Any other length is **irrational.**

It is worth remarking that Euclid's definition of (positive) rational numbers is equivalent to the one with which we have been working. If a number is the ratio of two natural numbers, say a/b, then it is rational because a/b and 1 are both measurable by the number $1/b$. Conversely, the only numbers that measure 1 are of the form $1/n$ where n is a natural number. Thus a number is commensurable with 1 only if it is of the form $m(1/n)$ where m is also a natural number.

Any two rational numbers a/b and c/d are both measurable by $1/bd$. Although it is impossible for a rational and an irrational to be commensurable, two irrational numbers can be commensurable. For example, $\sqrt{45}$ and $\sqrt{20}$ are measurable by $\sqrt{5}$.

Euclid also provided a method for determining whether two given numbers are commensurable. The largest number that measures two natural numbers is their greatest common divisor. A generalization of the Euclidean Algorithm makes it possible to find the "largest measure" of any two numbers *if it exists*. The following theorem is essentially Proposition II of Book X of the *Elements*.

[1] T. L. Heath, *A History of Greek Mathematics* (Oxford: The Clarendon Press, 1921), I, 403.

THEOREM | Let s and t be two lengths with t the longer. If the largest possible multiple of s is subtracted from t leaving r_1, and then the largest possible multiple of r_1 is subtracted from s leaving r_2, and then the largest multiple of r_2 is subtracted from r_1 leaving r_3, and this procedure continues without ever having any remainder measure the preceding one, then s and t are incommensurable. If a remainder measures the preceding remainder, then it also measures both s and t.

The theorem, stated in modern notation, says that if the natural numbers $n_1, n_2, \ldots$ and "numbers" $r_1, r_2, \ldots$ are found such that

$$\begin{aligned} t &= n_1 s + r_1 & (0 \leqslant r_1 < s) \\ s &= n_2 r_1 + r_2 & (0 \leqslant r_2 < r_1) \\ r_1 &= n_3 r_2 + r_3 & (0 \leqslant r_3 < r_2) \\ &\vdots & \vdots \end{aligned}$$

and no step that has a zero remainder is ever reached, then s and t are incommensurable. If at some step a zero remainder is obtained, then the last nonzero remainder measures all the other remainders, s, and t. (It will, in fact, be the largest number that does so.)

Example: Let us see what the algorithm yields for the numbers $\frac{5}{62}$ and $\frac{3}{14}$, even though we know they are commensurable and it is obvious that $1/(62)(14)$ will measure them. In order to facilitate computation we first write these fractions with a common denominator. We can then, essentially, apply the Euclidean Algorithm to the numerators.

$$\frac{5}{62} = \frac{5 \cdot 14}{62 \cdot 14} = \frac{70}{868} \qquad \frac{3}{14} = \frac{3 \cdot 62}{14 \cdot 62} = \frac{186}{868}$$

$$\frac{186}{868} = 2\left(\frac{70}{868}\right) + \frac{46}{868}$$

$$\frac{70}{868} = 1\left(\frac{46}{868}\right) + \frac{24}{868}$$

$$\frac{46}{868} = 1\left(\frac{24}{868}\right) + \frac{22}{868}$$

$$\frac{24}{868} = 1\left(\frac{22}{868}\right) + \frac{2}{868}$$

$$\frac{22}{868} = 11\left(\frac{2}{868}\right)$$

The "greatest common measure" of $\frac{5}{62}$ and $\frac{3}{14}$ is $\frac{2}{868}$ or $\frac{1}{434}$.

In general, the algorithm will yield $1/n$ where n is the least commo multiple of the denominators. The fact that any two rationals are com mensurable is what makes it possible to add them by finding a commo denominator.

Example: The array below shows the result of applying the algorithm i the less trivial case of the numbers $\sqrt{5}$ and 1. (There is a trick to actuall finding the required quotients.)

$$\begin{aligned}
\sqrt{5} &= 2 \cdot 1 + (\sqrt{5} - 2) \\
1 &= 4(\sqrt{5} - 2) + (\sqrt{5} - 2)^2 \\
(\sqrt{5} - 2) &= 4(\sqrt{5} - 2)^2 + (\sqrt{5} - 2)^3 \\
&\vdots \\
(\sqrt{5} - 2)^n &= 4(\sqrt{5} - 2)^{n+1} + (\sqrt{5} - 2)^{n+2} \\
&\vdots
\end{aligned}$$

Since we believe that a power of $(\sqrt{5} - 2)$ can never be zero, these com putations indicate that $\sqrt{5}$ is irrational because it is not commensurabl with 1.

To conclude this section we consider how the theorem could b proved. It is an algorithm, like the Euclidean Algorithm, that provides step-by-step reduction of the size of the numbers involved. A number wi measure s and t if and only if it measures the divisor and remainder i each step. (See the argument on p. 107, and check this statement on th array immediately following the theorem.) If some remainder measures th preceding remainder, then the algorithm terminates. The numbers s an t will both be measurable by that last remainder.

It remains to show that if the algorithm never provides a zer remainder, then the numbers must be incommensurable. When s and t we natural numbers, we used the Well-Ordering Axiom to show that since th remainders are natural numbers, the process must terminate. The W.O.A is not true for the set of all numbers that are lengths.

The remainders in the algorithm clearly become smaller and smalle In fact, since the remainder is smaller than the divisor, it must be less tha half of the number being divided. (Think about it!) Clearly, if the proce is continued long enough, then the remainders will become smaller tha any preassigned number. (This last statement is actually a form of th *Archimedean property* of the real numbers which is stated in the exercis of Chapters 15 and 16.)

Suppose that the algorithm does not terminate when applied to the commensurable numbers s and t. Let u be the number that measures s and t. Then each remainder is also measured by u. But eventually the remainders become smaller than u. Since it is impossible for u to measure a number smaller than itself, we have achieved a contradiction and proved the theorem.

Summary

The Pythagoreans proved that the lengths of some constructible line segments are not commensurable with the lengths of the segments used in the construction. They refused to accept irrational quantities as numbers. Greek mathematicians later proved theorems and developed techniques that are correct whether or not the quantities involved are rational.

Euclid defined the terms *rational* and *irrational* in terms of commensurability. He showed how a generalization of the Euclidean Algorithm can be used to determine whether two given numbers are commensurable.

EXERCISES

1. Show that the following pairs of numbers are commensurable by using Euclid's method to find the largest number that measures them.

(a) 15, 36 (b) $\frac{1}{15}, \frac{1}{36}$ (c) $\frac{1}{4}, \frac{1}{7}$

2. Prove that if p is a prime, then $\sqrt{p}$ is irrational. (This can be done by modifying the Pythagorean proof that shows that $\sqrt{2}$ is irrational.)

DISCUSSION QUESTIONS

1. Adapt the method used in Exercise 2 to prove that $\sqrt{6}$ is not rational. Generalize as much as possible.

2. It is a "well-known" fact that the decimal expansion of a number is finite or repeating if and only if the number is rational. Explain why a rational number must have such an expansion. You may find it helpful to consider the procedure for finding the expansion of rationals such as $\frac{1}{2}, \frac{1}{3}, \frac{1}{4}, \frac{1}{5}, \ldots$ Find a method for converting such a decimal expansion into the ratio of two natural numbers.

fourteen

approximation of irrational numbers

We have seen that irrational lengths exist. For practical applications, however, a good rational approximation of an irrational quantity will always suffice since the accuracy of a measurement is limited by the imperfections in the measuring instruments. For example, 1.4142 or 1.414 or maybe even 1.41 is as good an approximation of $\sqrt{2}$ as is usually needed in practice. It is intuitively clear that given an irrational length there is a rational length that approximates it to any desired degree of accuracy. This chapter deals with methods of finding such approximations.

The Babylonians knew the Pythagorean Theorem and therefore knew that the ratio of the diagonal of any square to the side is $\sqrt{2}$. They had estimates for this quantity. (See Discussion Questions of Chapter 1.) The Pythagoreans developed an algorithm for finding approximations as close as one pleases to $\sqrt{2}$. They also proved that the algorithm works.

Although the Pythagorean procedure is crude by present standards it is of historical interest because it shows the advantages of the Greek method of interpreting arithmetic in terms of geometry. Moreover, the algorithm has the elegance of a truly ingenious idea. We shall investigate the algorithm in detail and compare it with two more modern methods.

The computations involved in finding the approximations of $\sqrt{2}$ by the Pythagorean method are straightforward. We compute two sequences of numbers $d_1, d_2, d_3, \ldots$ and $s_1, s_2, s_3, \ldots$. The approximations are the quotients $d_1/s_1, d_1/s_2, \ldots$. The numbers d_1 and s_1 are both 1. Each of the other d's and s's is computed from the preceding ones by the formulas

$$s_{n+1} = s_n + d_n \qquad d_{n+1} = 2s_n + d_n.$$

That is, $s_2 = d_1 + s_1 = 2$, $d_2 = 2s_1 + d_1 = 3$, $d_3 = s_2 + d_2 = 5$, etc. The table below shows some of the approximations that result.

s	d	d/s
1	1	1
2	3	1.5
5	7	1.4
12	17	1.416...
29	41	1.4137...
70	99	1.41428...
169	239	1.414201...
408	577	1.4142107...
⋮	⋮	⋮

The algorithm can be justified by a geometric argument. The desired number is the ratio of the diagonal of any square to the side of the square. Consider the diagram below.

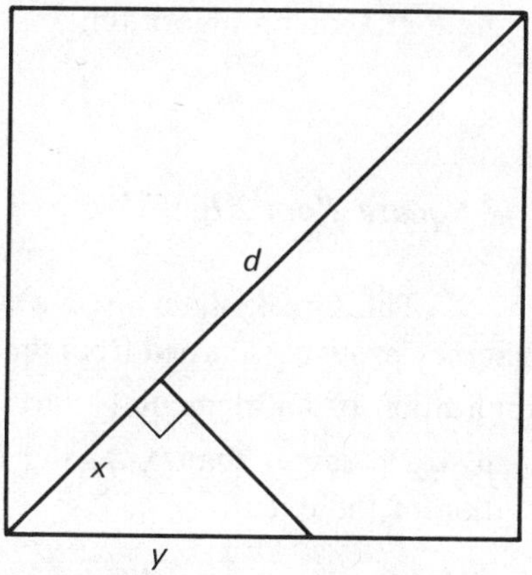

If the length of the side is laid off on the diagonal and a perpendicular is drawn to the diagonal, then the smaller triangle is similar to the larger one and is therefore half of a square. The ratio y/x must be equal to $\sqrt{2}$. This procedure could be continued, making smaller and smaller triangles, until the ratio could not be visually distinguished from 1.

The amount of error made in assuming that the diagonal of a very small square is the same as the side will be very small. If the relationship between the lengths in successive squares can be found, then we have an algorithm (by working backward) for finding an approximation of $\sqrt{2}$.

$$x = d - s = \sqrt{2}s - s, \qquad y = \sqrt{2}x = 2s - \sqrt{2}s.$$

Therefore we have

$$x + y = s \qquad 2x + y = d.$$

These equations show how to use the diagonal and side of a given square to compute the diagonal and side of the next larger square. They are exactly the equations used to define the d's and s's for the approximations. The assumption made in the algorithm is that the side and diagonal of the small square both have length 1. (The common value is chosen to be 1 because it simplifies the work. The choice makes no difference in the results because the approximations are ratios.)

This algorithm has at least two undesirable qualities. In the first place, it can only be used to approximate $\sqrt{2}$. Secondly, it takes a large number of steps to obtain a good estimate. In other words, the algorithm is neither generally applicable nor efficient.

We shall consider next two algorithms that do not have these deficiencies. The first is a special case of a method that has many practical applications and is very efficient. The second procedure is a generalization of the Pythagorean algorithm.

The Square Root Algorithm

The Greeks later used a method for approximating square roots that they probably learned from the Babylonians. It is based on the repeated application of an elementary fact: If a given number a is less than $\sqrt{A}$, then A/a is larger than $\sqrt{A}$, and conversely. This follows from a consideration of the equation

$$A = \sqrt{A} \cdot \sqrt{A} = a\left(\frac{A}{a}\right).$$

Square Root Algorithm: To approximate the square root of the rational number A, choose as the first approximation the largest integer a_1 such that $a_1^2 \leqslant A$. Compute the $(n + 1)$st approximation from the nth approximation by using the formula

$$a_{n+1} = \frac{1}{2}\left(a_n + \frac{A}{a_n}\right).$$

The method of the Square Root Algorithm is to choose as a_{n+1} the average of a_n and A/a_n. Since one of these numbers is larger than $\sqrt{A}$ and the other is smaller, the average will be a better approximation than a_n.

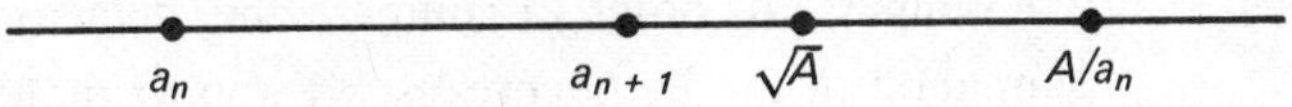

Example: We shall find a few approximations for $\sqrt{5}$. Since 5 is between 2^2 and 3^2, we choose $a_1 = 2$. Then

$$a_2 = \frac{1}{2}\left(a_1 + \frac{5}{a_1}\right) = \frac{1}{2}\left(2 + \frac{5}{2}\right) = \frac{9}{4} = 2.25$$

$$a_3 = \frac{1}{2}\left(a_2 + \frac{5}{a_2}\right) = \frac{1}{2}\left(\frac{9}{4} + \frac{20}{9}\right) = \frac{161}{72} = 2.2361 \ldots.$$

The approximation a_3 is accurate to the number of digits shown. If we desired better estimates of $\sqrt{5}$, we need only calculate $a_4, a_5, \ldots$.

Example: When the algorithm is applied with $A = 3$, we have $a_1 = 1$ and

$$a_2 = \frac{1}{2}\left(a_1 + \frac{3}{a_1}\right) = 2$$

$$a_3 = \frac{1}{2}\left(a_2 + \frac{3}{a_2}\right) = \frac{7}{4}$$

$$a_4 = \frac{1}{2}\left(a_3 + \frac{3}{a_3}\right) = \frac{97}{56}.$$

The value of $\sqrt{3}$ to four decimal places is 1.7321 and $\frac{97}{56}$ is 1.73214 ....

The condition that a_1 be chosen as in the algorithm is just a convenient way of ensuring that the first approximation is good enough to make the algorithm work. In practice one would choose a_1 to be as good an approximation as possible.

It is not difficult to assess the efficiency of this algorithm and at the same time show that an approximation to any desired degree of accuracy can be obtained. The error made by using the nth approximation is $a_n - \sqrt{A}$. A little algebraic manipulation will yield the inequality

$$a_{n+1} - \sqrt{A} < (a_n - \sqrt{A})^2.$$

That is, the error made by using the approximation a_{n+1} is less than the square of the error made by using the preceding approximation.

If, for example, the error made by a_n is known to be less than 0.01, then the error made by a_{n+1} is less than 0.0001 and the next approximation will be within 0.00000001 of the correct value. This algorithm provides good approximations very quickly.

Example: In order to compare the efficiency of this algorithm with the method of the Pythagoreans, we approximate $\sqrt{2}$. The results are

$$a_1 = 1 \qquad a_3 = \frac{1}{2}\left(\frac{3}{2} + \frac{4}{3}\right) = \frac{17}{12}$$

$$a_2 = \frac{1}{2}(1 + 2) = \frac{3}{2} \qquad a_4 = \frac{1}{2}\left(\frac{17}{12} + \frac{24}{17}\right) = \frac{577}{408}$$

It takes two steps of the Pythagorean method to achieve the improvement in the estimate made between a_2 and a_3, and it takes four steps more to achieve the estimate a_4. The differences become even more pronounced if the comparison is carried further. Since the estimate $\frac{3}{2}$ is within 0.1 of $\sqrt{2}$, the approximation $\frac{577}{408}$ is accurate to *within* 0.0001. Furthermore if we were to calculate a_7, then we would have an estimate of $\sqrt{2}$, which makes an error of less than 0.1 to the power 2^5. This bound on the error is 1 with 31 zeros in front of it.

Continued Fractions

The method used by the Pythagoreans to estimate $\sqrt{2}$ is a special case of a general procedure for obtaining approximations by using the Euclidean Algorithm. Euclid showed how the algorithm can be used to determine whether or not two numbers are commensurable. If a zero remainder is eventually obtained when the algorithm is applied to the numbers 1 and x, then x is rational. If the remainders become smaller and smaller but none is ever zero, then x is irrational.

When x is irrational, the quotients and remainders calculated in applying the algorithm can be used to find rational approximations of x. The procedure is to assume that one of the small remainders is zero and evaluate the consequences of that assumption. The assumption that a specific remainder is zero is equivalent to the assumption that x is a particular rational number. If the remainder chosen is very small, then the rational number will be a good approximation of x.

Example: The array obtained when the Euclidean Algorithm is applied to $\sqrt{2}$ and 1 is as follows:

$$\begin{aligned}
\sqrt{2} &= 1 \cdot 1 + (\sqrt{2} - 1) \\
1 &= 2(\sqrt{2} - 1) + (\sqrt{2} - 1)^2 \\
\sqrt{2} - 1 &= 2(\sqrt{2} - 1)^2 + (\sqrt{2} - 1)^3 \\
&\vdots \\
(\sqrt{2} - 1)^{n-1} &= 2(\sqrt{2} - 1)^n + (\sqrt{2} - 1)^{n+1} \\
&\vdots
\end{aligned}$$

Since the remainders are powers of $(\sqrt{2} - 1)$ and none is zero, the array shows that $\sqrt{2}$ is irrational. If the first remainder, $(\sqrt{2} - 1)$, is supposed equal to zero, we have the estimate

$$\sqrt{2} \approx 1 \cdot 1 = 1.$$

A better estimate is obtained if the second remainder, $(\sqrt{2} - 1)^2$, is set equal to zero. The second equation then becomes $1 = 2(\sqrt{2} - 1)$ or $(\sqrt{2} - 1) = \frac{1}{2}$. If this value of the first remainder is substituted into the first equation, we obtain

$$\sqrt{2} \approx 1 + 1/2 = 3/2.$$

The estimates that result from assuming that r_3 is zero and r_4 is zero are

$$1 + \frac{1}{2 + \frac{1}{2}} = \frac{7}{5} \qquad 1 + \cfrac{1}{2 + \cfrac{1}{2 + \frac{1}{2}}} = \frac{17}{12}.$$

It will be helpful to introduce some new terminology and notation.

DEFINITION | If $a_1, a_2, \ldots, a_n$ are natural numbers, then the quantity

$$a_1 + \cfrac{1}{a_2 + \cfrac{1}{a_3 + \cfrac{\vdots}{\cfrac{1}{a_{n-1} + \cfrac{1}{a_n}}}}}$$

is called a **continued fraction** and is denoted $[a_1, a_2, \ldots, a_n]$.

The approximations of $\sqrt{2}$ are the continued fractions, [1], [1, 2], [1, 2, 2], [1, 2, 2, 2], ... These are exactly the same estimates that were obtained by the Pythagorean method. Although that method could be used only to find estimates for $\sqrt{2}$, the Euclidean Algorithm can, theoretically, be applied to obtain approximations for any irrational number. In order to estimate the number x, the algorithm is applied to the numbers x and 1. If the quotients in the array are $n_1, n_2, n_3, \ldots$ (see p. 126), then the continued fraction approximations will be $[n_1]$, $[n_1, n_2]$, $[n_1, n_2, n_3]$, ... Unfortunately, there is no simple—or even moderately difficult—way to actually calculate the quotients for most irrational numbers.

Example: The result of applying the Euclidean Algorithm to $\sqrt{5}$ and 1 is shown on p. 126. Thus, the continued fraction approximations for $\sqrt{5}$ are

$$[2], \qquad [2, 4] = \tfrac{9}{4}, \qquad [2, 4, 4] = \tfrac{38}{17}, \qquad [2, 4, 4, 4, \ldots].$$

It was not until the sixteenth or seventeenth century that this general method of obtaining approximations was developed. Leonhard Euler (1707–1783) stated the modern version of the method in a paper entitled *Continued Fractions.* Euler was the greatest algorist and the most prolific mathematician that has ever lived. (He also had thirteen children and was known for his powers of concentration.) Euler was born in Switzerland but he spent most of his professional life at the Academies founded by Catherine I in St. Petersberg and Frederick I in Berlin. Despite the fact that he was blind for many of his last years, Euler produced immense amounts of important mathematics.

Although the method of obtaining continued fraction approximations for an irrational number from the Euclidean Algorithm is more difficult and less efficient than other methods, the estimates that are found this way have several important properties. Two of the significant characteristics of continued fractions are listed below.

1. The continued fraction approximations for irrational square roots are periodic, that is, the numbers $a_1, a_2, a_3, \ldots$ will behave like "repeating" decimal expansions for rational numbers. For example, the approximations for $\sqrt{3}$ are of the form [1, 1, 2, 1, 2, 1, 2, ...] and those for $\sqrt{10}$ are [3, 6, 6, 6, ...]. Of course, if continued fraction estimates are found for the rational number m/k by applying the algorithm to m/k and 1, then the process will be finite and produce an equality:

$$m/k = [a_1, a_2, a_3, \ldots, a_n].$$

2. The approximations obtained by using continued fractions in this

Leonhard Euler. *Euler was the greatest mathematician of his time. His most influential work was a book which compiled and organized the techniques of the calculus. He invented much of the notation which is now used in mathematics.* (*George Arents Research Library*)

way are in one sense the best possible estimates. If $p_n/q_n = [a_1, a_2, \ldots, a_n]$ is the nth such rational number approximation for an irrational number α, then p_n/q_n is a better approximation for α than any other fraction with denominator less than or equal to q_n. It is known, further, that the error made by p_n/q_n is less than $1/q_nq_{n+1}$. The computations for $\sqrt{2}$ show that the best estimate with a denominator less than or equal to 70 is $\frac{99}{70}$ and $\frac{99}{70}$ is accurate to within $1/(70)(169)$.

The nature of the mathematician's interest in algorithms such as these should perhaps be made more clear. It is of interest to find methods by which rational approximations to any desired degree of accuracy can be found for irrational numbers. It is also of interest to have procedures for deciding how large the possible error can be at any given step. If estimates are actually needed, it is desirable that the algorithm be efficient and that it be possible to program a machine to carry out the computations. A mathematician is not so much interested in finding an estimate for $\sqrt{2}$ that is accurate to a thousand decimal places as he is interested in being able to prove that it can be done and in knowing the best way to do it.

Summary

Our intuitive beliefs about the properties of irrational numbers suggest that it should be possible to approximate any irrational as closely as one wishes with rational numbers. The continued fractions constructed from numbers acquired by applying the generalized Euclidean Algorithm provide such approximations. The Square Root Algorithm and similar methods are practical procedures for actually computing rational estimates for many irrational numbers.

EXERCISES

1. Use the Square Root Algorithm to find approximations for the square roots of 6, 8, 10, and 50. In each case, stop when the fractions become complicated. For reference, six-digit approximations of these numbers are

$$\sqrt{6} \approx 2.44949 \qquad \sqrt{10} \approx 3.16228$$
$$\sqrt{8} \approx 2.82843 \qquad \sqrt{50} \approx 7.07107$$

2. Confirm that $[2, 3, 4] = \frac{30}{13}$ and $[3, 5, 7] = \frac{115}{36}$ by changing the continued fractions into quotients of natural numbers. Show that these continued fractions for $\frac{30}{13}$ and $\frac{115}{36}$ can be obtained by using the Euclidean Algorithm.

3. Use the continued fraction expansions below to obtain a few decimal-fraction approximations for $\sqrt{6}$, $\sqrt{8}$, $\sqrt{10}$, $\sqrt{50}$.

$$\sqrt{6} = [2, 2, 4, 2, 4, \ldots] \qquad \sqrt{10} = [3, 6, 6, 6, \ldots]$$
$$\sqrt{8} = [2, 1, 4, 1, 4, \ldots] \qquad \sqrt{50} = [7, 14, 14, 14, \ldots]$$

DISCUSSION QUESTIONS

1. Try to modify the Square Root Algorithm so that it can be used to approximate cube roots and other roots.

2. It is easy to find the rational number that has a given repeating decimal expansion. Try to find a method for discovering the number that has a given repeating continued fraction expansion. In particular, try to identify [1, 1, 1, 1, . . .].

fifteen

the real number system: construction of the rationals

Up to this point our discussion of numbers and arithmetic has been true to the spirit of the times in which the concepts were developed. Although we have made a distinction—as did the Greeks—between things which are commensurable with 1 and those which are not, we have in general simply accepted and used the numbers and the rules for their manipulation. This is consistent with the attitudes of most of the mathematicians who lived before 1800.

As mentioned in the discussion of geometries, a reexamination of the fundamental concepts of mathematics was begun in the nineteenth century because of logical contradictions and uncertainties that had been found in the many new mathematical systems developed since the seventeenth century. Attempts were made to find axioms from which all mathematics, arithmetic as well as geometry, could be deduced. The concepts and the structures which are discussed informally in this chapter are representative of those which were developed from about 1870 to 1920, using the experience and knowledge gained over a period of two thousand years, and which most mathematicians find logically sound by today's standards.

The system that we are to discuss is called the *real number system*. The basic idea is that every kind of number we have discussed is to be a real number. The problem is to choose axioms and definitions so that the resulting system will conform to what we want real numbers to be. We would, in addition, like the system to have essentially just the one model that we expect it to have. Once we see that this can indeed be done, then we can again dispense with formalities and work with the numbers according to our "well-known" rules. If challenged to produce reasons or proof for doing things the way we do, we can refer to the axioms of the system and demonstrate the validity of our methods.

Giuseppe Peano (1858–1932) was professor of "infinitesimal analysis" at Turin University. One of his major accomplishments was a formal development of all mathematics. In 1908 he published his *Formulario Mathematico,* which summarizes his careful, precise reconstruction of mathematics.

The first section of the book is devoted to mathematical logic and set theory. Peano introduced there the undefined symbols and the axioms for working with them that he used in the following sections to develop mathematics. In this section of the book he gave a careful discussion of the conventions of language and the rules for working with sets of objects. Much of the notation he introduced has been adopted for general use. He used, for example, the symbols $\in$, $\cup$, and $\cap$ to denote *is an element of, the union of sets,* and *the intersection of two sets,* respectively. The following sections were devoted to Arithmetic, Algebra, Geometry, Limits, Differential Calculus, and Integral Calculus in that order.

The value of Peano's formulation of the foundations of mathematics was recognized, but it was not widely accepted in its entirety. Mathematicians of the twentieth century have continued the attempts to make mathematics free of logical defects.

The Natural Numbers

The obvious place to start with the formal construction of the axiomatic system of numbers and their arithmetic is with the natural numbers. What is needed is a list of axioms from which all the familiar properties of natural numbers can be developed. The axioms used below are called Peano's Postulates and are the ones used in the *Formulario Mathematico.* (Peano is not, however, responsible for the informality of this version.)

Besides assuming the basic facts about set theory and logic, Peano chose exactly three undefined terms to be used in arithmetic. N represents a set with elements that are called *natural numbers.* "1" is an object called *one.* (Peano actually used 0, which he called zero, and made zero a natural number. It really is only a matter of terminology and makes no difference at all. We shall conform to common practice and exclude 0.) Furthermore, there is a relationship denoted " $'$ ". If a is an element of N, then a' is called the *successor of* a. There are then five axioms.

The Peano Postulates

1. 1 is in N.
2. If a is in N, then a' is in N.
3. If a is in N, then $a' \neq 1$.
4. If a and b are in N and $a' = b'$, then $a = b$.
5. If M is a subset of N, then $M = N$ provided
   (a.) 1 is in M
   (b.) If a is in M, then a' is in M.

Since the entire real number system and all its properties ar deduced from these axioms, it is worthwhile to consider their properties They were chosen so that they would have the "well-known" countin numbers as a model. The set of axioms should therefore be consistent.

The first axiom is independent because without it the set N coul be assumed to be empty. The model in which $N = \{1\}$ and $1' = 2$ satisfie all the axioms except the second one. It follows from the second axiom tha there must be a "chain" of numbers in the set. The third axiom provide that 1 be the first element in the chain.

The model shown in the diagram on the left below, in which ar arrow is drawn from each number to its successor, shows that the third axion is independent. It may seem at first glance that the second and third axiom taken together imply that there are an infinite number of elements in N The diagram on the right shows that this is not the case. This model show that the fourth axiom is independent, and necessary if N is not to contair a finite number of elements by "doubling back."

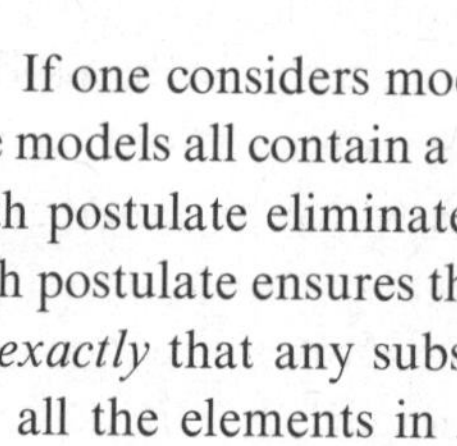

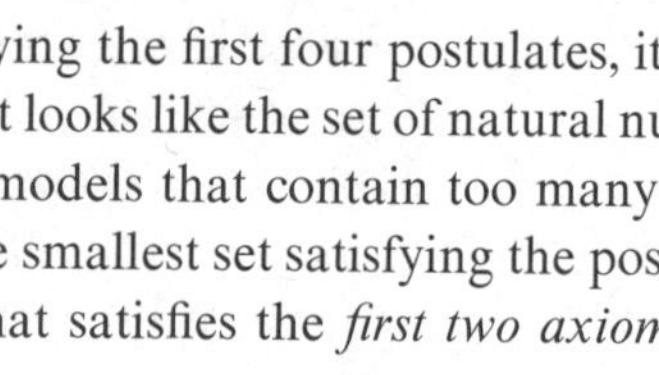

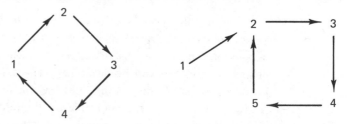

If one considers models satisfying the first four postulates, it is seen that the models all contain a subset that looks like the set of natural numbers. The fifth postulate eliminates all the models that contain too many things. The fifth postulate ensures that N is the smallest set satisfying the postulates. It says *exactly* that any subset of N that satisfies the *first two axioms* must contain all the elements in N.

The true test of the usefulness of Peano's axioms comes when the system is developed by making definitions and proving theorems. If we accept these postulates, then the next thing to do is define addition, multiplication, and the inequality $<$ on N and prove properties such as the associative, commutative, and distributive laws. This project is beyond the scope of this course, not because of its difficulty but because of its length. (It would be worth your time to at least browse through the complete development which can be found, for example, in the books by Landau and Evans which are included in the bibliography.)

We shall only consider briefly how the definitions could be made.

DEFINITION | If m and n are natural numbers, then $n + m$ is defined by

1. $n + 1 = n'$
2. $n + m' = (n + m)'$.

DEFINITION | If m and n are natural numbers, then $n \cdot m$ is defined by

1. $n \cdot 1 = n$
2. $n \cdot m' = n \cdot m + n$.

DEFINITION | If m and n are natural numbers, then m is less than n, written $m < n$, when there is a natural number k such that $m + k = n$.

Notice that the fifth postulate, which is often called the *Axiom of Mathematical Induction* (A.M.I.), ensures that these definitions will define addition and multiplication for all natural numbers. The A.M.I. is a very strong assumption and is the principal tool used to prove many of the basic theorems about the natural numbers.

Let $P(n)$ be a symbol standing for a statement that mentions the natural number n. Suppose that you wish to prove as a theorem "$P(n)$ is true for all natural numbers n." If M is the set of natural numbers for which $P(n)$ is true, then the theorem is proved if M satisfies the two conditions of the A.M.I.

The A.M.I. is actually equivalent to the Well-Ordering Axiom (see p. 106). The W.O.A. can be stated more simply and is more easily understood than the A.M.I. It would have been used in place of the A.M.I. except that to do so one would have to explain somehow the meaning of the word "smallest" in its statement. The concept of size in the set of natural numbers depends on the definition of the inequality $<$. It is more convenient to use Peano's axioms and prove the W.O.A. as a theorem.

Example: To see how the A.M.I. can be applied and compare its application with that of the W.O.A., two proofs of the theorem "$2^n > n$ for all natural numbers n" are given below.

Proof 1: Consider the set A of all numbers for which the inequality is *not* true. If A is not empty, then by the W.O.A. it must contain a smallest number x. Clearly x is not 1 because $2^1 > 1$ and 1 is *not* in A. Thus x is greater than 1 and $x - 1$ is a natural number that is *not* in A. Therefore, $2^{x-1} > x - 1$. But then

$$2 \cdot 2^{x-1} > 2(x - 1) \quad \text{and} \quad 2^x > x + (x - 2) \geqslant x$$

so x is *not* in A. The assumption that x is the smallest element in A leads to the conclusion that x is not in A. This is a contradiction. A must be *empty*.

Proof 2: Consider the set M of numbers for which the inequality *is* true. Clearly 1 is in M because $2^1 > 1$. Suppose that the number x *is* in M. Thus, $2^x > x$. We have then $2^{x+1} > 2x$ and $2^{x+1} > x + x \geqslant x + 1$. Therefore, $x + 1$ is in M if x is in M. The set M satisfies the conditions of the A.M.I. and must contain all natural numbers.

The Rational Numbers

Once the properties of the natural numbers have been established, we can expand the set of numbers by making definitions. As with the natural numbers, our choice of definitions will be determined by our desire to conform to "reality."

Since we have a pretty good idea of what we want, it seems reasonable to next define the positive rational numbers. These numbers are ratios of natural numbers. They are pairs of natural numbers in which it makes a difference which number occurs first. This suggests the following informal versions of the usual definitions.

DEFINITION | A **positive rational number** is an ordered pair, written (a, b) or a/b, of natural numbers a and b.

DEFINITION | If a/b and c/d are positive rational numbers, we define the sum and product, denoted by symbols $\oplus$ and $\odot$, as

$$\frac{a}{b} \oplus \frac{c}{d} = \frac{ad + bc}{bd}$$

$$\frac{a}{b} \odot \frac{c}{d} = \frac{ac}{bd}.$$

After the properties of positive rational numbers are established, the next step is to define the negative rational numbers and zero. Remembering that subtraction bears the same relation to addition as division does to multiplication, it is natural to define rational numbers as ordered pairs of positive rationals. Again ignoring the finer points we could make the following definitions.

DEFINITION | A **rational number** is an ordered pair, (r, s) or $r - s$, of positive rational numbers.

Example: By definition the pairs $(\frac{1}{4}, \frac{3}{4})$, $(7, 5)$, $(3, 9)$, and $(\frac{1}{3}, \frac{1}{3})$ are rational numbers. In practice it is immediately shown that

$$(\tfrac{1}{4}, \tfrac{3}{4}) = (0, \tfrac{1}{2}) \qquad (3, 9) = (0,6)$$
$$(7, 5) = (2, 0) \qquad (\tfrac{1}{3}, \tfrac{1}{3}) = (1, 1)$$

and these numbers are abbreviated $-\frac{1}{2}$, 2, -6, and 0.

The motivation for the following definitions of sum and product should become clear upon computing a few examples.

DEFINITION | If (r, s) and (t, u) are rational numbers, we define

$$(r, s) \boxplus (t, u) = (r \oplus t, s \oplus u)$$
$$(r, s) \boxdot (t, u) = [(r \odot t) \oplus (s \odot u), (r \odot u) \oplus (s \odot t)].$$

This short discussion should be enough to convince you that the rational numbers can be developed axiomatically. That is, axioms can be found, appropriate definitions of the operations can be made, and the usual properties can be proved as theorems if one has the desire and patience to do so. It is then possible to be sure that a "well-known" fact about the rational numbers is as valid as the axioms for the system and the logic used in proving the fact as a theorem.

Summary

About one hundred years ago the foundations were laid for the careful axiomatic development of mathematics. A theory of sets was constructed first and the formal structure of the other branches of mathematics was based upon set theory. This is one of the reasons for the emphasis on sets in the modern mathematics curricula.

A set of axioms for the real number system was discovered by Peano. These axioms give explicit assumptions about the set of natural numbers. The rest of the rational numbers can be defined, in stages, as special sets of natural numbers. The usual rules for arithmetic can be proved as theorems in this system.

EXERCISES

1. Use the definitions of addition and multiplication of positive rational numbers to compute

$$\tfrac{3}{4} + \tfrac{5}{6} \qquad \tfrac{7}{8} + \tfrac{2}{1}$$
$$(\tfrac{3}{4})(\tfrac{5}{6}), \qquad (\tfrac{7}{8})(\tfrac{2}{1}).$$

2. How should the relation $<$ be defined for positive rational numbers?

3. Use the definitions for addition and multiplication of rational numbers to compute

$$(3 - 5) + (4 - 1) \qquad (3 - 5)\cdot(4 - 1)$$
$$(\tfrac{1}{2} - \tfrac{1}{4}) + (\tfrac{1}{6} - \tfrac{1}{6}) \qquad (\tfrac{1}{2} - \tfrac{1}{4})\cdot(\tfrac{1}{6} - \tfrac{1}{6})$$

Replace each of the expressions in this exercise by the symbol that is commonly used, and check your answers.

4. Prove each of the following statements by using the A.M.I. (This is a subtle method and requires careful thought. You will probably find it useful to write an outline of what must be done.)

(a) The number $n(n + 1)$ is divisible by 2 for every natural number n.

(b) If x is a fixed positive number, then $(1 + x)^n \geqslant 1 + nx$ for every natural number n.

(c) For every value of n, the sum of the first n even natural numbers is $n(n + 1)$.

DISCUSSION QUESTIONS

1. Prove each of the statements of Exercise 4 by using the Well-Ordering Axiom. Compare the general form of the two methods of proof.

***2.** Prove the Archimedean Property of the set of rational numbers: If r and s are rational numbers, then there exists a natural number n so that $rn > s$. (*Hint*. Use the W.O.A.) Find an informal interpretation of this property.

sixteen

the real number system: cuts and sequences

In this section we complete our survey of how the real number system can be developed axiomatically. We have seen how the rational numbers can be defined. The goal now is to extend the system of rational numbers to a system that includes the irrational quantities we have used in the preceding chapters.

There are two basically different ways to achieve this goal. It is possible to list a set of axioms that will completely determine the system we want. It would be difficult, however, to understand the reasons for the choice of the axioms that are used without having a clear idea of what the fundamental properties of the numbers should be. We shall therefore use the other method first. We shall assume the rational numbers and their properties to be known and define the real numbers in terms of the rational numbers.

In our informal discussions of real numbers we assumed the following two statements about the properties of these numbers.

1. For every length there is a number, either rational or irrational, and conversely.

2. Any irrational number can be approximated by a rational with an error as small as desired.

The formal definition of the real number system should be made in such a way that it is possible to prove these two statements as theorems. In fact, each of the two common methods of defining the real numbers in terms of rationals was developed because of the belief that one of the statements characterizes the set of real numbers. One of these distinct, bu equivalent, definitions was used first by J. W. R. Dedekind and the othe is the work of Georg Cantor.

Dedekind Cuts

J. W. Richard Dedekind (1831–1916) received his Ph.D. fror Göttingen when Gauss was director of the observatory there. In spite c the fact that he spent most of his life (50 years) teaching in a secondar school, Dedekind made many important contributions to the theory of num bers. He was, for example, the first to give the definition of "infinite" tha is used today: "A set is infinite if there exists a one-to-one correspondenc between the set and one of its proper subsets."

Dedekind believed that the essential property of the real numbe is that there are exactly enough numbers to measure any line segment. H saw this as a consequence of the assumption that there is a one-to-on correspondence between real numbers and the points on a line. On the othe hand he felt that it was important that the ideas of "line" and/or "length"– geometric concepts—should not be part of the foundations upon which th real number system is constructed. Dedekind preferred to base the rea number system on the theory of sets.

Dedekind's actual definition has the simplicity of genius. He imag ined a line on which a point had been labeled 0 and every other rationa number had been assigned a point at an appropriate distance from 0. Ther are then some points that have not been labeled with numbers.

Dedekind thought of a real number as a *cut* of this line. Any cu divides the line into two pieces. These pieces can be described complete by listing the rational numbers associated with each one. Dedekind define real numbers as partitions of the set of rational numbers.

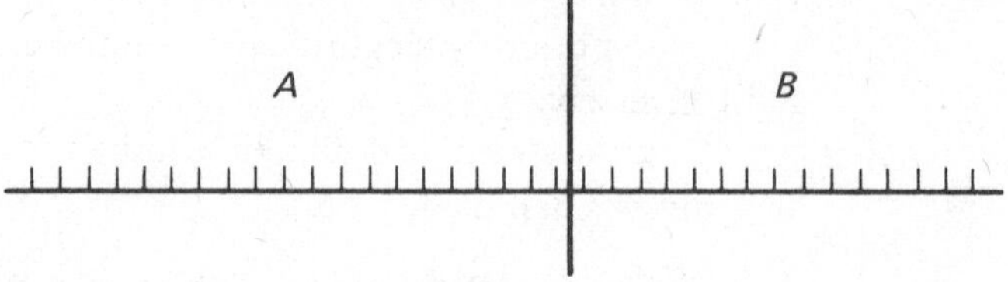

DEFINITION (Dedekind) | A *real number* is a pair of sets (A, B) of rational numbers such that

1. Neither A nor B is empty and every rational number is in one of the sets.

2. If r is in A and s is in B, then $r < s$.

Since we want to be able to think of rational numbers as being real numbers, it is desirable to have some way of associating each rational number with one of Dedekind's real numbers. If the cut occurs at a rational number, then it is natural to call the resulting real number a rational real number and identify with it that rational number.

DEFINITION | A real number (A, B) such that A has a largest member or such that B has a smallest member is called a **rational** real number. Otherwise (A, B) is an **irrational** real number.

Example: The following sets show how some particular numbers can be defined by Dedekind's method. The set given is the "left" set and the "right" set is assumed to be all other rational numbers.

0 : {all negative rational numbers}
$\sqrt{2}$: {all rational numbers x such that $x < 0$ or $x^2 < 2$}
π : {all rational numbers less than the ratio of the circumference of a circle to its diamter}

The real number zero is obviously a rational real beacuse the "right" set contains a smallest element, namely the rational number 0. It is not difficult to show that $\sqrt{2}$ is irrational because neither set of the cut that defines it has an "end" number. It is very difficult to prove that π is irrational.

Using this definition of real number as a special kind of pair of sets of rational numbers, it is possible to prove that the set of real numbers has all the desired properties.

In particular, it can be proved that there is a one-to-one correspondence between the set of real numbers and the set of points on a line. Since geometric concepts are excluded from the formal development, the theorem actually states a property of sets of real numbers. Suppose that the set of all real numbers is partitioned into subsets A and B in the same way that cuts of rational numbers are defined. Then it is necessary and sufficient to show that "this cut of the real number line occurs at a real number." In terms of sets this means that either A or B has an "end" element. Dedekind's Hauptsatz (Principal Theorem) says exactly that.

THEOREM | Let A and B be subsets of the set of real numbers such that

1. Neither set is empty and every real number is in one of the sets
2. If x is in A and y is in B, then $x < y$.

Then either A has a largest element or B has a smallest element.

A set of numbers that has the property described in the theorem is said to be *complete*. This theorem is often referred to as the *Completeness Theorem* for the set of real numbers. Notice that the set of rational numbers is not complete. The conclusion of the theorem is false when A and B are sets of rational numbers such that (A, B) is an irrational real number.

Cantor Sequences

A second method of defining the real numbers from the rational numbers was invented by Georg Cantor (1845–1918). Cantor was one of the pioneers in the study of the theory of sets and a personal friend of Dedekind. The work that he did with infinite sets is a masterpiece that shows the power of reason in areas where intuition is untrustworthy.

Cantor's work was ridiculed by many of the prominent mathematicians of his time. It is thought that one of them, Leopold Kronecker, was responsible for the fact that Cantor was never able to obtain a position at the prestigious University of Berlin. Kronecker was sort of a "Latter Day Pythagorean" and completely distrusted the use of infinite quantities in mathematics. He is remembered best for his statement "God made the whole numbers, all the rest is the work of man."

Although in the last years of his life Cantor was given the recognition due to him, the criticism he received had contributed to several nervous breakdowns and he spent his last days in a mental hospital.

Cantor's definition is based upon the observation that any irrational number should be approximable by rational numbers. His idea was that a real number is completely determined by a sequence of its rational approximations. In Cantor's development of the system a real number is defined by a special kind of sequence of rational numbers. The conditions on the sequence ensure that the numbers of the sequence get closer and closer together without having to say that there is something to which they get close.

DEFINITION (Cantor) | A **real number** is a sequence $r_1, r_2, r_3, \ldots$ of rational numbers such that for every rational number ϵ, no matter how small, there is a natural number k such that $|r_n - r_m| < \epsilon$ when m and n are greater than k.

Example: More than one sequence will define the same real number. Any rational real number r is defined by either of the sequences below.

$$r, r, r, r, \ldots \qquad r - 1, r - \tfrac{1}{2}, r - \tfrac{1}{3}, r - \tfrac{1}{4}, \ldots$$

An irrational real number is defined by its sequence of decimal fraction approximations or by sequences obtained by algorithms such as those in Chapter 14. For example, $\sqrt{2}$ is determined by any one of the sequences

$$1, 1.4, 1.41, 1.414, \ldots \qquad 1, \tfrac{3}{2}, \tfrac{7}{5}, \tfrac{17}{12}, \ldots$$
$$[1], [1, 2], [1, 2, 2], \ldots \qquad a_1, a_2, a_3, \ldots$$

where $a_1 = 1$ and $a_{n+1} = \frac{1}{2}(a_n + 2/a_n)$.

Technically these sequences are not the same real number. They are equal if equality is defined. In the same sense $\frac{1}{2}$ and $\frac{2}{4}$ are equal positive rational numbers and $3 - 5$ and $4 - 6$ are equal negative numbers but they are not identical.

The set of real numbers obtained by Cantor's method has exactly the same properties as Dedekind's set of real numbers. For each theorem in one of these systems there is a corresponding theorem in the other. Cantor's version of the completeness theorem says that if a sequence of *real* numbers satisfies the condition of his definition, then they get very close to a real number. In other words, any number approximable with real numbers is itself a real number. Of course the formal statement of the theorem expresses these concepts very precisely.

Axioms for the Real Number System

As mentioned at the beginning of this chapter, it is also possible to define the real number system all at once by listing sufficient postulates. Since we now have a better idea of what the axioms should be, such a definition follows.

DEFINITION | The **real number system** is a set R of elements $0, 1, x, y, z, \ldots$ with the operations of multiplication and addition and the relation $<$ defined on R such that the following conditions are satisfied.

1. R is closed under addition. (If x and y are in R, then $x + y$ is in R.)

2. R is closed under multiplication.

3. Addition is commutative. ($x + y = y + x$ for all real numbers x and y.)

4. Multiplication is commutative.

5. Addition is associative. ($x + (y + z) = (x + y) + z$.)

6. Multiplication is associative.

7. Multiplication distributes over addition. ($x(y + z) = (xy) +$ (xz).)

8. The number 0 is an additive identity. ($x + 0 = x$.)

9. The number 1 is a multiplicative identity.

10. Each real number has an additive inverse. (For each x there is an $\bar{x}$ such that $x + \bar{x} = 0$.)

11. Each real number, except 0, has a multiplicative inverse.

12. For each real number x, exactly one of the following is true: $x < 0$, $x = 0$, $0 < x$.

13. If $0 < x$ and $0 < y$, then $0 < x + y$ and $0 < x \cdot y$.

14. If S is a nonempty subset of R that is bounded above, then S has a smallest upper bound.

It is of little interest, at this point, whether these fourteen axioms are independent or not. They were chosen because they provide a convenient, simply understood summary of the properties of the real number system from which other facts can be proved without too much difficulty. This definition of the real numbers, rather than that of Cantor or Dedekind, is the "working definition" used when the numbers themselves are not the main subject of investigation.

Summary

Cantor and Dedekind gave definitions of real numbers. They both defined a real number as a special kind of set of rational numbers and each proved that the resulting set of real numbers has the desired properties.

The real number system can also be defined as the set with operations that satisfy a list of conditions. The sets of real numbers defined by Dedekind and Cantor both satisfy this definition.

DISCUSSION QUESTIONS

1. (a) How can sums and products of real numbers be defined when the cut and sequence definitions of real numbers are used?

(b) How can the relation $<$ be defined in each case?

(c) Find cuts and sequences that can be identified as 0, 1, and -1.

*(d) Check your answers to the questions above by making sure that your model satisfies the set of axioms for R.

2. Investigate the significance of condition 14 of the definition of R. Compare this condition with the completeness results of Dedekind and Cantor. Show that the set of rational numbers does not satisfy this condition

3. Assume that the real numbers are defined by the list of axioms of this chapter and that no definitions have ever been given for the natural numbers, integers, or rational numbers. Define these sets of numbers as subsets of the real number system by selecting those properties of the reals that are satisfied and making new conditions as necessary.

***4.** Explain how one might prove the Archimedean property of the real numbers: If x and y are positive real numbers, then there exists a natural number n such $nx > y$.

part four

algebra

seventeen

theory of equations

Now that we have seen the basis for the rules for working with the real numbers we shall consider the application of these rules. The word algebra is a corruption of the Arabic "al-jabr" which was used by the ninth century mathematician Mohammed al-Kwārizmī in the title of one of his works. He used the word to refer to some of the manipulations involved in transforming an equation like $x^2 + 3x = 10$, which gives an implicit description of the real number x, into the form $x = 2$, which describes x explicitly. During the Middle Ages the mathematicians of Europe adopted the word *algebra* to mean the study of methods of finding the numbers which satisfy conditions which are expressible as equations. This subject is now more accurately called the *theory of equations* and the word *algebra* has a variety of more general meanings.

In this chapter we shall consider methods of finding the real numbers that satisfy equations involving the sums and products of powers of the number. Any problem of this type can be written in the form

$$a_n x^n + a_{n-1} x^{n-1} + \cdots + a_1 x + a_0 = 0$$

where x is the number or numbers sought and the numbers $a_0, a_1, \ldots a_n$, which are called coefficients, are specified by the conditions of the problem. This type of notation is the result of a long period of experimentation to find a way to represent the problems in a form that is compact and easy to work with. We shall use simplified variations of this notation when convenient. The expression on the left is called a **polynomial of degree *n*** and the entire expression is called a **polynomial equation.** The polynomials of degrees 1, 2, 3, and 4 are usually referred to as linear, quadratic, cubic, and quartic, respectively.

Linear Equations

Even very early records of mathematical activity contain problems that require the solution of a linear equation. Although a general method was never stated explicitly, a procedure for solving any linear equation involving positive rational numbers was known four thousand years ago.

Example: One of the oldest examples of the solution of a linear equation occurs as Problem 27 in the Rhind papyrus.[1] Slightly simplified it is

A quantity and its $\frac{1}{5}$ added together become 21. What is the quantity? Assume 5. Since $5 + \frac{1}{5}(5) = 6$, as many times as 6 must be multiplied to get 21, so must 5 be multiplied to give the quantity.

\1	6	\1	5
\2	12	\2	10
\\$\frac{1}{2}$	3	\\$\frac{1}{2}$	$2\frac{1}{2}$
Total	21	Total	$17\frac{1}{2}$

The method used here is to make a convenient guess at the answer and then see what relation the guess has to the correct answer. (A convenient guess in this case is one that makes computation easy.) This problem, in modern notation, is *if* $x + \frac{1}{5}x = 21$, *what is* x?

The solution to any linear equation is easily found using modern techniques. Any linear equation can be written in the form $ax + b = 0$. This kind of equation has exactly one solution and it is $-b/a$. For example, the solution of $15x - 8 = 0$ is $\frac{8}{15}$.

The solution of $ax + b = 0$ was found by subtracting b from both sides of the equation and then dividing by a. The procedure, without the modern notation, appears in the work of al-Kwārizmī, but it was not until several hundred years later that it was realized and accepted that negative solutions could have a meaningful interpretation.

Quadratic Equations

Although the Egyptians could solve some problems involving quadratic equations, they apparently had no general method. The Baby-

[1] A. B. Chace, *et al.*, *The Rhind Papyrus* (Oberlin, Ohio: M.A.A., 1927–1929), two volumes.

lonians knew a procedure that is essentially the same as the one we use today. This method, which is called "completing the square," was also used by the Hindus and Arabs. The Greeks used a geometric procedure for constructing a line segment that has length equal to the solution. The Greeks also proved geometrically that the Babylonian method works.

Example: The following is a problem like those found on Babylonian cuneiform tablets.

What is the side of a square if the area is $3\frac{3}{4}$ more than the side? Divide 1, the coefficient, into two parts. $\frac{1}{2}$ times $\frac{1}{2}$ is $\frac{1}{4}$. Add $\frac{1}{4}$ to $3\frac{3}{4}$. 4 has the root 2. Add to 2 the $\frac{1}{2}$. The side of the square is $2\frac{1}{2}$.

In modern notation the problem and solution are

$$x^2 - x = 3\tfrac{3}{4}, \qquad x^2 - x + \tfrac{1}{4} = 4, \qquad (x - \tfrac{1}{2})^2 = 4,$$
$$x - \tfrac{1}{2} = 2, \qquad x = 2\tfrac{1}{2}.$$

The Babylonians used only positive numbers and thus found only one square root for each number.

Example: A typical, easy problem as stated by a ninth- or tenth-century Arab mathematician might have read as follows.

A square and 8 roots are equal to 65 units. Now the roots in this problem are 8. Therefore, take half of 8, which squared is 16, and add to 65. This gives 81, which has square root 9. Subtract half the roots, leaving 5.

In modern notation the problem and solution are

$$x^2 + 8x = 65, \qquad x^2 + 8x + 16 = 81, \qquad (x + 4)^2 = 9^2,$$
$$x + 4 = 9, \qquad x = 5.$$

The Arabs also ignored the possibility that a number might have a negative square root.

The medieval Europeans learned much algebra from the Arabs. (The *Algebra* of abū Kāmil, which is listed in the bibliography, was translated into English from a Hebrew translation of the Arabic that was made in the fourteenth or fifteenth century by a Spanish Jew named Mordecai Finzi.) The Arabs had preserved many of the Greek methods and results. They often used geometric figures to justify their results.

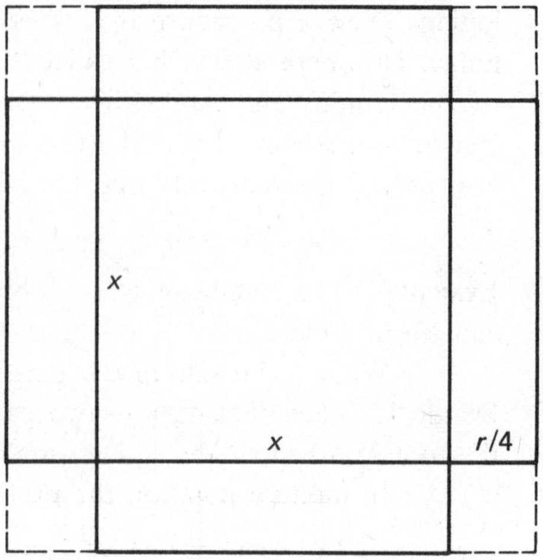

The technique of adding "the square of half the roots" to obtain a number that is a perfect square was explained with a diagram like that above. The "plus" shape has an area of $x^2 + rx$. The sum of the areas of the four corner pieces is $4(r/4)^2$, which is equal to $(r/2)^2$. When these are added to the "plus," the result is a square with area $(x + r/2)^2$. In modern notation this is expressed as

$$x^2 + rx + \left(\frac{r}{2}\right)^2 = \left(x + \frac{r}{2}\right)^2.$$

The procedure used today for solving quadratic equations is exactly the same as the one above. Our notation and acceptance of the existence of all real numbers make it possible to state a general method.

Any quadratic equation can be written in the form $ax^2 + bx + c = 0$. When the procedure is applied, the result is formulas, giving two values of x as solutions.

$$x^2 + \left(\frac{b}{a}\right)x + \frac{c}{a} = 0$$

$$x^2 + \left(\frac{b}{a}\right)x + \left(\frac{b}{2a}\right)^2 = \left(\frac{b}{2a}\right)^2 - \frac{c}{a}$$

$$\left(x + \frac{b}{2a}\right)^2 = \left(\frac{b}{2a}\right)^2 - \frac{c}{a}$$

$$x + \frac{b}{2a} = \sqrt{\left(\frac{b}{2a}\right)^2 - \frac{c}{a}} \qquad x + \frac{b}{2a} = -\sqrt{\left(\frac{b}{2a}\right)^2 - \frac{c}{a}}$$

$$x = \frac{-b + \sqrt{b^2 - 4ac}}{2a} \qquad x = \frac{-b - \sqrt{b^2 - 4ac}}{2a}$$

Examples: The solutions of $x^2 - 14x + 33 = 0$ are

$$\frac{14 + \sqrt{64}}{2} = \frac{14 + 8}{2} = 11 \qquad \frac{14 - \sqrt{64}}{2} = 3$$

The solutions of $3x^2 + 8x - 2 = 0$ are

$$\frac{-8 + \sqrt{88}}{6} = \frac{-4 + \sqrt{22}}{3} \qquad \frac{-8 - \sqrt{88}}{6} = \frac{-4 - \sqrt{22}}{3}$$

The formulas will provide two different real solutions whenever $b^2 - 4ac$ is positive and one (or two identical) real solution(s) when $b^2 - 4ac = 0$. If $b^2 - 4ac$ is negative, the formulas contain the square root of a negative number. Such quantities are not real numbers and no real numbers are solutions of the corresponding equations.

Equations of Higher Degree

Although Archimedes had a geometric technique for solving some cubic equations and his method was improved by Omar Khayyam about 1100 A.D., there were few successful attempts at solving problems involving cubed quantities before the Renaissance. The adoption of Arabic symbols for numbers made it possible for European mathematicians to handle problems that they had previously found too cumbersome. The modern algebraic notation, such as the use of letters to denote unknown numbers, the method of writing exponents, and the use of symbols for the operations, was not developed until the seventeenth century. The first method that could be used to solve any cubic equation was discovered in the sixteenth century, in spite of the difficulties involved in stating and explaining the solution.

A general method for solving cubic equations was probably known by Scipione del Ferro (1465–1526). The procedure was rediscovered before 1540 by Niccolo Tartaglia (1500?–1557) and published by him, in verse form, in 1546; however, his method was first made public in 1545. Geronimo Cardano (1501–1576) learned the procedure from Tartaglia after pledging to keep it secret. Cardano then published his book on algebra entitled *The Great Art* and included the explanation of how to solve cubic equations.

Although he had earned a medical degree by the time he was 25, Cardano was not granted a license to practice for many years. (As a young man he published a book entitled *Book on Poor Methods of Healing in Present Use*.) Cardano supported himself after he graduated from college

Niccolo Tartaglia. *Tartaglia's actual family name was Fontana; the word "tartaglia" is a nickname meaning "stutterer" which he was given and used. Tartaglia taught mathematics in Verona, Brescia, and Venice and wrote a book on gunnery entitled "A New Science."* (*George Arents Research Library*)

by lecturing on mathematics and medicine. He is credited with formulating some of the basic rules of probability and he wrote several popular books on healing and astrology. Cardano achieved fame in his later years as a physician and was called to attend the royal courts in many parts of Europe.

Cardano's method for solving cubic equations is based on a series of algebraic manipulations that reduce the problem to a quadratic equation. The solutions are given in the form of complicated combinations of the coefficients of the cubic polynomial. The calculation of the solutions from these formulas requires finding the cube roots of quantities which look like the formulas which give the solutions of quadratic equations.

When given a cubic equation $ax^3 + bx^2 + cx + d = 0$, the first step in the procedure is to divide by a. Then some manipulations are performed that change the equation to the form $x^3 + px + q = 0$, where p and q are combinations of a, b, c, and d. After a few pages of computations the result is

$$x = \sqrt[3]{-\frac{q}{2} + \sqrt{\frac{q^2}{4} + \frac{p^3}{27}}} + \sqrt[3]{-\frac{q}{2} - \sqrt{\frac{q^2}{4} + \frac{p^3}{27}}}.$$

This method appears to provide just one solution to every cubic equation. Cardano was concerned, moreover, about the fact that in some problems, for which he knew the answers, the method seemed to yield "meaningless" expressions for the solutions.

Examples: A few numerical examples will demonstrate that Cardano's method provides answers that are difficult to deal with.

The equation $x^3 + 3x - 14 = 0$ clearly has a solution $x = 2$. The formula yields

$$x = \sqrt[3]{7 + \sqrt{50}} + \sqrt[3]{7 - \sqrt{50}}.$$

It is not obvious whether or not this number is really equal to 2. The equation $x^3 - 33x - 18 = 0$ has the solution $x = 6$, but the formula gives

$$x = \sqrt[3]{9 + \sqrt{-1250}} + \sqrt[3]{9 - \sqrt{-1250}}.$$

This is not only complicated but also calls for the square root of a negative number.

The computations are clearly very involved and of little consequence here. What it is important to note is that a method had been found that could be used to find formulas for the solutions of any cubic equation and

Geronimo Cardano. *A prominent Italian mathematician, physician, and astrologer, Cardano contributed to the advance of methods for solving equations. Cardano was an inveterate gambler and discovered several of the basic laws of probability as a consequence of his studies of games of chance.* (*George Arents Research Library*)

that the procedure involves only the operations of addition, subtraction, multiplication, division, and taking roots.

Even though there were some difficulties in interpreting the answers that Cardano's method provided, the problem of finding a general method for solving cubic equations was regarded as answered and attention was turned to the next most difficult type of equation. (The ambiguities in Cardano's solution were finally eliminated by Euler in 1732.)

Ludovic Ferrari (1522–1565), a student of Cardano, found a similar method for solving every quartic equation. He discovered a technique that will always reduce a quartic equation to a cubic equation. This makes finding the solutions of a quartic equation even more complicated than finding the solutions of a cubic equation, but it can be done with the same simple algebraic operations. Mathematicians immediately began to try to adapt the procedure to solve equations of higher degrees. The methods that were tried in their attempts to reduce the general equation of degree five to an equation of degree four, so that it could be solved by Ferrari's method, resulted instead in equations of degree *six*. The search for an extension of the method that would provide solutions to the general equation of degree higher than four was unsuccessful.

Summary

Until the middle of the nineteenth century, algebra was the study of methods of solving polynomial equations. The techniques necessary to solve linear and quadratic equations have been known for thousands of years. Procedures that can be used to solve all cubic and quartic equations were developed in the sixteenth century.

Since the same technique can be applied to find the solutions of every polynomial of a given degree (less than five), it is possible to write formulas expressing the solutions in terms of the coefficients. These expressions consist of sums, differences, products, quotients, and roots of the coefficients. Formulas of this type for the solutions of equations of degree higher than four do not exist. (This subject is discussed further in Chapter 20.)

EXERCISES

1. Solve each of the following equations.

$$3x + 2 = 0 \qquad -x + 4 = 0 \qquad 5x - 3 = 0$$

2. Graph the curves of each of the following equations.

$$y = 3x + 2 \qquad y = -x + 4 \qquad y = 5x - 3$$

Check that the solutions to Exercise 1 are the first coordinates of the point where these curves cross the x-axis.

3. Finding the real numbers that satisfy the equation $ax^2 + bx + c = 0$ can be interpreted geometrically as finding the points where the parabola $y = ax^2 + bx + c$ intersects the line $y = 0$ (the x-axis). Check this by solving the following equations and sketching the corresponding parabolas.

$$x^2 - 2x - 8 = 0$$
$$2x^2 - 12x + 18 = 0$$
$$x^2 + 1 = 0$$

DISCUSSION QUESTIONS

1. Use the geometric interpretation and what you have learned about solving quadratic equations to formulate rules for determining from the coefficients a, b, and c whether the quadratic equation $ax^2 + bx + c = 0$ has

(a) No real solutions.
(b) One real solution.
(c) Two real solutions, both positive.
(d) Two real solutions, both negative.
(e) Two real solutions, one positive and one negative.

Be able to give a geometric interpretation of each case.

***2.** Must a cubic equation have at least one real solution? Investigate to see what the graphs of some specific cubic polynomials look like. Can you suggest a method for finding crude approximations to those solutions that exist? What can be said about the solutions of equations of higher degrees?

eighteen

approximate solutions of equations

We have seen that methods exist that can be used to find all solutions of polynomial equations of degree less than or equal to four. These algorithms work no matter what the coefficients are and provide the exact solutions as algebraic combinations of the coefficients. It is desirable to have techniques for finding the solutions to equations of any degree. Furthermore, the formulas for the solutions of cubic and quartic equations are not very practical. They involve extensive complicated computations, the results are difficult to translate into an understandable form, and they are of the "all or nothing at all" variety. If a trivial error is made in the calculations, the numbers obtained will probably not even be close to the true values of the solutions.

It would suffice in most cases to have a practical method for finding arbitrarily good approximations to the solutions of equations. In this chapter we shall investigate some algorithms that can be used for this purpose.

Bounds for Solutions

In order to use an algorithm to find approximations for the solutions of an equation it is necessary to have, to begin with, some information about the value of the solutions to be estimated. The methods that we shall discuss require that bounds for a solution be known; that is, two numbers must be provided such that the solution is larger than one of the numbers and smaller than the other. These bounds, in a sense, constitute the first approximations of the solution. The bounds can be found by using the following theorem.

THEOREM | If the value of the polynomial $p(x) = a_nx^n + a_{n-1}x^{n-1} + \cdots + a_0$ is positive when $x = s$ and negative when $x = t$, then it is zero for some real number between s and t.

For example, the cubic equation $5x^3 + 4x^2 + x - 7 = 0$ has a solution between 0 and 1 because in this case $p(0) = -7$ and $p(1) = 3$. The equation $x^2 - 3 = 0$ has the expected solution between 1 and 2 because $p(1) = -2$ and $p(2) = 1$. It is easily verified that there is another solution between -1 and -2.

The proof of this theorem depends on the completeness property of the real numbers. We shall not investigate the proof; the geometric interpretation is convincing. Consider the graph of the curve $y = p(x)$. The theorem says that if the curve is below the x-axis at s and above the axis at t, then the curve crosses the axis between s and t. The x-coordinate of the point of intersection is a solution of $p(x) = 0$.

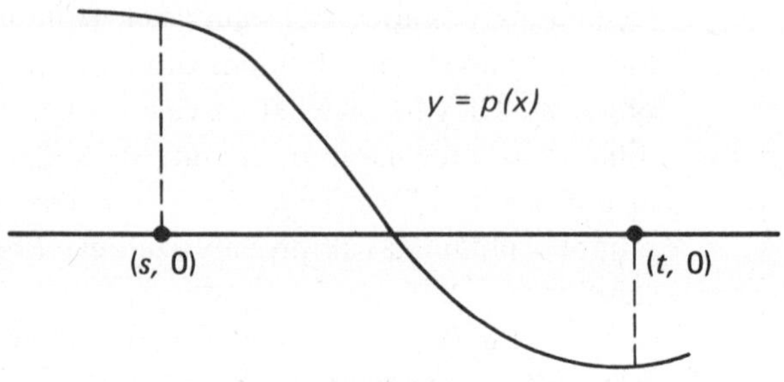

The information provided by the theorem can be systematically applied to provide sequences of approximations to the solution lying between the bounds. There are a number of algorithms that can be used to improve the bounds and provide approximations arbitrarily close to the solution.

Suppose that $p(x)$ is a polynomial that has a solution between s and t. An approximation of the solution that is accurate to within any number ϵ can be found by calculating $p(s + \epsilon)$, $p(s + 2\epsilon)$, $p(s + 3\epsilon)$, ... or $p(t - \epsilon)$, $p(t - 2\epsilon)$, ... and continuing until the value of the polynomial changes sign. When this occurs, each of the new bounds is an approximation of the desired accuracy. This method will always work but is, of course, grossly inefficient. The application of a little common sense makes this kind of approach more practical.

Example: Suppose that we wish to find a real solution of $2x^3 - 27x^2 + 4x - 54 = 0$. We first determine which multiples of 10 the solution lies between. A moderate amount of computation yields $p(0) = -54$, $p(10) = -714$, $p(20) = 5226$. There must be a solution between 10 and 20. Since -714 is closer to zero than is 5226, the solution is probably closer to 10. If we compute more values, we find $p(13)$ is -171 and $p(14)$ is 198. There should be a solution about halfway between 13 and 14. Further computation reveals that 13.5 is a solution.

Thirteenth-century Chinese mathematicians combined the intelligent use of this trial and error approach with a technique that simplifies the computations involved to obtain an effective method for solving polynomial equations. By the fifteenth century, Arab mathematicians were using a similar procedure. The procedure is now known as *Horner's Method* because it came into widespread usage in Europe only after a paper explaining the method was published by W. G. Horner (1773–1827).

If it is known that a solution to an equation lies between two specific numbers, then Horner's Method can be used to find the exact decimal value of the solution if the solution is rational and an arbitrarily good decimal approximation if the solution is irrational. The method has the advantage that it automatically provides information about the accuracy of each approximation.

The Method of Double False Position

The second algorithm that we shall consider is sometimes called the *method of double false position*. The "false positions" referred to are bounds for the solution. The method of double false position is an algorithm based on a specific way of choosing the next approximation (or bound, if you prefer) between the given bounds (approximations). Each new approximation is used with the appropriate one of the preceding bounds to form the basis for a further estimation.

The motivation for the method of finding the approximation and the means of actually computing it are most easily explained with the use of analytic geometry.

Let $p(x)$ be a polynomial and let x_1 and x_2 be bounds for a solution of $p(x) = 0$. The points $(x_1, p(x_1))$ and $(x_2, p(x_2))$ are on the curve $y = p(x)$. The straight line through these two points will divide the interval between x_1 and x_2 on the x-axis into pieces proportional to the sizes of $p(x_1)$ and $p(x_2)$. The x-coordinate of the point of intersection of this line with the x-axis is chosen as the next approximation.

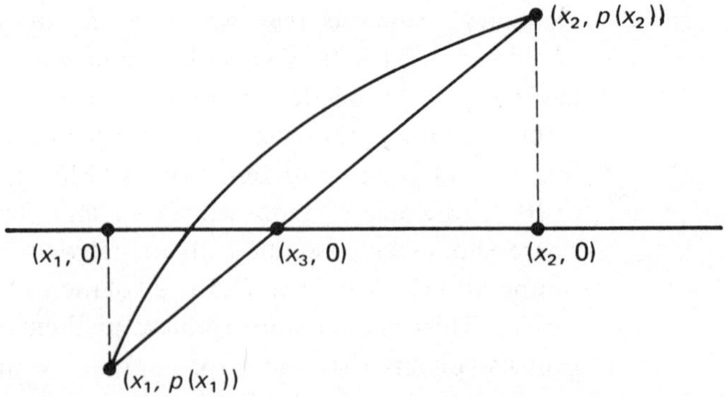

Since the application of this algorithm to other than the simplest equations tends to involve tedious computations and is best done with the aid of a computer, we shall illustrate the details on a trivial example and then show the geometric interpretations of the applications to more complicated problems.

Example: Let $p(x) = x^2 - 5$. Since $p(2) = -1$ and $p(3) = 4$, we can choose $x_1 = 2$ and $x_2 = 3$. The points on the curve are $(2, -1)$ and $(3, 4)$ and the line through the points has the equation

$$\frac{y+1}{x-2} = \frac{4+1}{3-2} = 5.$$

The intersection of this line with the x-axis occurs when $y = 0$. Letting $y = 0$ and solving for x, we obtain $x_3 = \frac{11}{5}$.

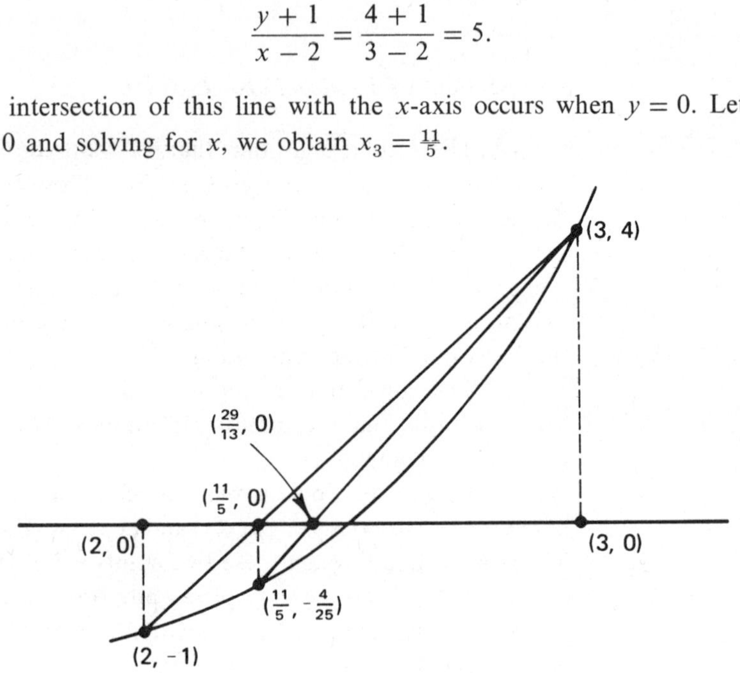

Since $p(\frac{11}{5}) = -\frac{4}{25}$ is negative, we find x_4 by using x_3 with x_2. The points are (3, 4) and $(\frac{11}{5}, -\frac{4}{25})$. The equation of the line joining them is

$$\frac{y-4}{x-3} = \frac{4+\frac{4}{25}}{3-\frac{11}{5}} = \frac{26}{5}.$$

Setting $y = 0$, we obtain $x_4 = \frac{29}{13}$. If a better approximation were desired, then x_4 would be used with x_2 because $p(x_4)$ is negative.

Examples: The results of applying the algorithm with initial approximations x_1 and x_2 and a variety of choices of $p(x)$ are shown graphically below.

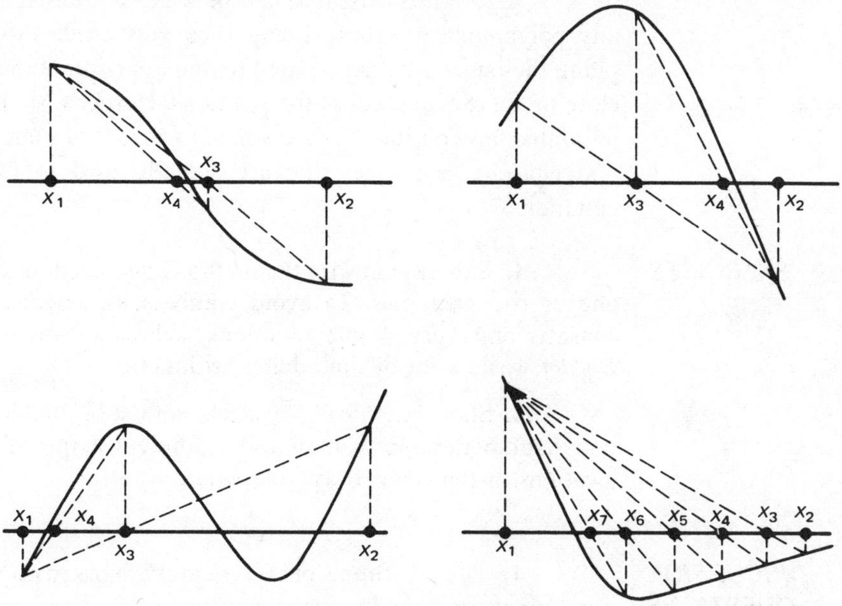

The method of double false position always *seems* to provide approximations that converge to a solution. It is not necessarily true, however, that the new estimate at a given step is better than both of those from which it is computed. Also, the efficiency of the method depends on the shape of the graph of the equation. The study of algorithms for finding numerical solutions to problems such as polynomial equations is the branch of mathematics called *numerical analysis*.

Some very good algorithms, which can be used to solve several different kinds of difficult problems in mathematics, have been known for more than two hundred years. Until recently many of these methods were impractical because of the enormous amount of computation necessary to actually apply them. They were of importance mainly because they were the bases for proofs that solutions do exist and can, theoretically at least, be found. There has been a revival of interest in finding new algorithms and improving old ones in the last quarter century because of the development of computers that can perform large numbers of calculations very rapidly.

Summary

Algorithms exist that can be used to approximate the solutions of any polynomial equation. Using even very crude estimates as a starting point, these methods can be used to find approximations that are arbitrarily close to the exact values of the solutions. Horner's Method and the method of double false position are elementary versions of some of the sophisticated and efficient techniques that are actually used to find the solutions of equations.

EXERCISES

1. Experiment with the methods discussed in this chapter by doing one or two examples. To avoid cumbersome computations it is best to consider only very simple equations such as $x^2 - 6 = 0$ and $x^3 - 5 = 0$. *Do not* waste a lot of time doing arithmetic.

2. Show graphically how the method of double false position gives better and better approximations for different shapes of curves and different locations of the original approximations.

DISCUSSION QUESTIONS

1. The algorithms of this chapter are based on methods of choosing new approximations between existing ones. Use your imagination and geometric considerations to invent a new algorithm by devising a "search pattern" for selecting a sequence of approximations. Try to evaluate the practicality of your new method.

2. Investigate whether and when the following method of "single false position" will work: It is assumed now that it is always possible to find the equation of the line tangent to any point of the graph of the equation. The method is to pick any "reasonable" first approximation x_1, then as the next approximation, choose the point of intersection of the tangent line through $(x_1, p(x_1))$ with the x-axis. Continue by finding x_3 using the tangent line at $(x_2, p(x_2))$. Be sure to check the method for a variety of shapes of curves and different locations of x_1.

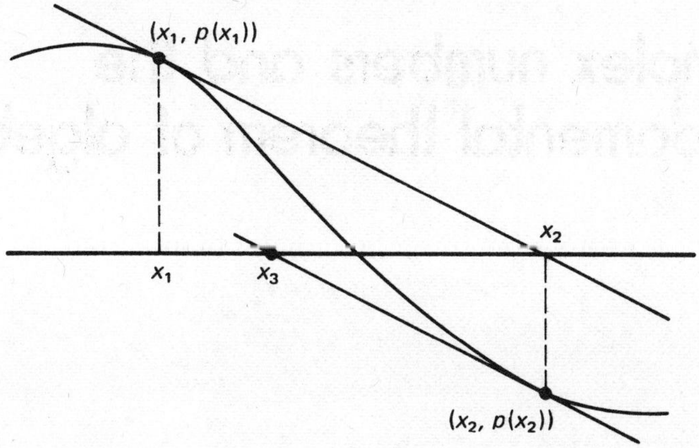
$(x_1, p(x_1))$
x_2
x_1
x_3
$(x_2, p(x_2))$

nineteen

complex numbers and the fundamental theorem of algebra

As noted in Chapter 17, the mathematicians of the sixteenth an seventeenth centuries sought formulas that would give the solutions o polynomial equations in terms of the coefficients of the polynomials. The found such formulas for polynomials of degree less than five but wer unsuccessful in attempts to use generalizations of their methods for findin such relationships in the cases of polynomials of higher degree. In thi chapter we shall consider some of the facts they discovered in the cours of their investigations.

The Fundamental Theorem of Algebra

Suppose that it is desired to find a polynomial equation that wil have specified solutions. The equation $x - r_1 = 0$ clearly has exactly on solution, $x = r_1$. The equation

$$(x - r_1)(x - r_2) = x^2 - (r_1 + r_2)x + r_1r_2 = 0$$

has two obvious solutions. Since the graph of the corresponding curve i a parabola, the solutions $x = r_1$ and $x = r_2$ should be the only ones.

Similarly, the equation $(x-r_1)(x - r_2)(x - r_3) = 0$ certainly ha at least three solutions. When the product is multiplied out, the equation i

$$x^3 - (r_1 + r_2 + r_3)x^2 + (r_1r_2 + r_1r_3 + r_2r_3)x - r_1r_2r_3 = 0.$$

At this point our geometric information is insufficient and it is not clea whether or not r_1, r_2, and r_3 are the only solutions.

Example: An equation with just two solutions, 4 and 5, is $(x - 4)(x - 5) = x^2 - 9x + 20 = 0$, where the -9 is $-(4 + 5)$ and $20 = 4 \cdot 5$. An equation with solutions at 2, -3, and 6 is

$$x^3 - (2 - 3 + 6)x^2 + (-6 + 12 - 18)x - (-36) = 0$$
$$x^3 - 5x^2 - 12x + 36 = 0.$$

It should be obvious that it is always possible to construct a poly nomial of degree n that will have any n numbers (at least) as solutions. Furthermore, there seems to be a very nice pattern that enables one to compute the coefficients of the polynomial if one knows the solutions. (Unfortunately the problem that the mathematicians were trying to solve is exactly the opposite of this.)

Consider again the problem of finding all solutions of a given equation. For the linear equation $ax + b = 0$, it is clear from geometrical consideration of the line $y = ax + b$ that there is exactly one solution. If the equation is written so that the "lead" coefficient is 1,

$$x + \frac{b}{a} = 0,$$

then this is of the form $x - r_1 = 0$ where r_1 is the solution. We saw above that if an equation with exactly one solution, r_1, was desired, then $x - r_1 = 0$ will do the job. Now, *conversely,* we see that all linear equations, if written so that the coefficient of x is 1, have exactly one solution and the solution is the negative of the constant term.

It has been noted that the equation $x^2 - (r_1 + r_2)x + r_1r_2 = 0$ has exactly the two solutions r_1 and r_2. The next question is whether every quadratic equation

$$ax^2 + bx + c = 0 \quad \text{or} \quad x^2 + \left(\frac{b}{a}\right)x + \frac{c}{a} = 0$$

has exactly two solutions such that the sum is $-b/a$ and the product is c/a. If it is assumed that every number has two square roots, then the quadratic formula yields

$$r_1 = \frac{-b + \sqrt{b^2 - 4ac}}{2a} \quad \text{and} \quad r_2 = \frac{-b - \sqrt{b^2 - 4ac}}{2a}.$$

If $b^2 - 4ac$ is positive, then these quantities are real numbers and the sum and product are indeed $-b/a$ and c/a. If $b^2 - 4ac = 0$, then $r_1 = r_2 = -b/2a$. If we interpret this as *two identical* solutions, then the formulas still hold because in this case

$$r_1 r_2 = \frac{b^2}{4a^2} = \frac{c}{a}.$$

If $b^2 - 4ac$ is negative, then r_1 and r_2 are not real numbers. The graph of the parabola $y = ax^2 + bx + c$ shows that there are no real solutions. These quantities involving the square roots of negative numbers were rejected as "absurd," "meaningless," "fictitious," and "imaginary" by mathematicians until the seventeenth century. At that time it was realized that if reasonable rules were made for adding and multiplying these quantities, then the formulas continue to be true. The sum of the roots is $-b/a$ and the product is c/a.

Rafael Bombelli (1526–1572) was an Italian engineer. He was known during his lifetime as an expert in the draining of marshes. Bombelli became interested in Cardano's *Ars Magna* and wrote a textbook on algebra in which he attempted to clarify the results about cubic and quartic equations. Bombelli invented symbols to express the quantities that Cardano had described with prose. Although he had doubts about the value of the idea, he also showed that an arithmetic of the roots of negative numbers can be defined in a manner consistent with the occurrence of these quantities as solutions of equations.

Cardano's formula for the solutions of a cubic equation seems to provide only one solution. In modern notation one of his results is

$$x = \sqrt[3]{-\frac{q}{2} + \sqrt{\frac{q^2}{4} + \frac{p^3}{27}}} + \sqrt[3]{-\frac{q}{2} - \sqrt{\frac{q^2}{4} + \frac{p^3}{27}}}.$$

Many cubic equations, however, were known to have three real solutions. The quantity in the formula is a sum of cube roots. If an interpretation is given which assigns 3 cube roots to every number, then it is possible that the formula for the coefficients in terms of the roots might (will) be true for all cubic equations.

Various schemes were devised for handling the roots of negative numbers, but most mathematicians remained convinced that such quantities had no meaning and no use. One of the first men to make significant use of them was Albert Girard (1595–1632). He discovered that if one accepted these numbers and made the natural rules for working with them, then the facts known about the solutions of equations made more sense and formed a very general pattern. In 1629 Girard published a book entitled (approximately) *A New Invention in Algebra*. In this work he stated some very important theorems.

FUNDAMENTAL THEOREM OF ALGEBRA | Every polynomial equation of degree n has exactly n solutions.

THEOREM | If the numbers $r_1, r_2, \ldots, r_n$ are the solutions of the equation $x^n + a_1x^{n-1} + \cdots + a_{n-1}x + a_n = 0$, then,

$$\begin{aligned} r_1 + r_2 + \cdots + r_n &= (-1)a_1 \\ r_1r_2 + r_1r_3 + \cdots + r_{n-1}r_n &= (-1)^2a_2 \\ &\vdots \\ r_1r_2r_3 \cdots r_n &= (-1)^na_n. \end{aligned}$$

The theorems are clearly not true unless "imaginary" numbers are accepted as solutions. The theorems cannot be proved unless these imaginary numbers are defined and shown to have the desired properties. For this reason, and because it is very difficult, the Fundamental Theorem of Algebra was not proved for almost two hundred years. The proof was given by Gauss in his doctoral dissertation in 1799.

Example: Whatever the solutions of $2x^4 + 5x^3 + 6x^2 - 8x + 7 = 0$ may be, the Fundamental Theorem of Algebra says that there are exactly four of them. If they are called r_1, r_2, r_3, and r_4, then Girard's theorem yields the following information:

$$\begin{aligned} r_1 + r_2 + r_3 + r_4 &= -\tfrac{5}{2} \\ r_1r_2 + r_1r_3 + r_1r_4 + r_2r_3 + r_2r_4 + r_3r_4 &= 3 \\ r_1r_2r_3 + r_1r_2r_4 + r_1r_3r_4 + r_2r_3r_4 &= 4 \\ r_1r_2r_3r_4 &= \tfrac{7}{2}. \end{aligned}$$

The Complex Numbers

In the end the controversy over imaginary numbers was resolved by the same method used in the cases of negative and irrational numbers. The method is to formally extend the number system to include such numbers.

The new system of numbers includes all numbers of the form $s + ti$ where s and t are real numbers and i is an abbreviation for $\sqrt{-1}$. This representation suggests that such numbers, which are called **complex numbers,** be defined formally as ordered pairs of real numbers. Such a definition was made, addition and multiplication were defined, and appropriate theo rems were proved about the behavior of complex numbers. Since at thi point you should be convinced that the formalities can be carried out i necessary, we shall simply state the definitions of addition and multiplicatio informally and consider briefly some of the properties of complex number (We shall not need to *use* complex numbers in later chapters.)

DEFINITION | The sum and product of the complex numbers $a + bi$ and $c + di$ are define by the equations

$$(a + bi) + (c + di) = (a + c) + (b + d)i$$
$$(a + bi) \cdot (c + di) = (ac - bd) + (ad + bc)i.$$

We have seen in the past that a geometric interpretation of algebrai quantities is often very useful. The fact that a complex number is define by a pair of real numbers suggests a procedure for displaying these numbe graphically. If the x-axis is interpreted as the "real" axis and the y-axis a the "imaginary" axis, then the complex number $r + si$ can be represente as the point (r, s).

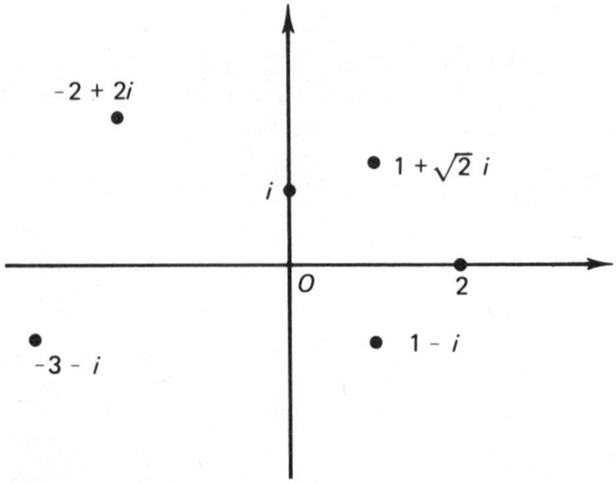

Such a representation is called an *Argand diagram* in honor of French mathematician, although it was first suggested by Caspar Wesse (1745–1818), a Norwegian. The significance of this "natural" geometri interpretation probably seems minor at this point. It should be remembere that at the time that Wessel published his results (1797) the use of "imag nary" quantities was not generally accepted, the Fundamental Theorem c Algebra had not been proved, and the formal development of a system c numbers was still nearly a hundred years in the future. The fact that thes numbers do have a geometric interpretation that is useful in describing physical phenomena was the most important reason for the decision to include them in the formal system of numbers.

The addition and multiplication of complex numbers can be depicted on Argand diagrams.

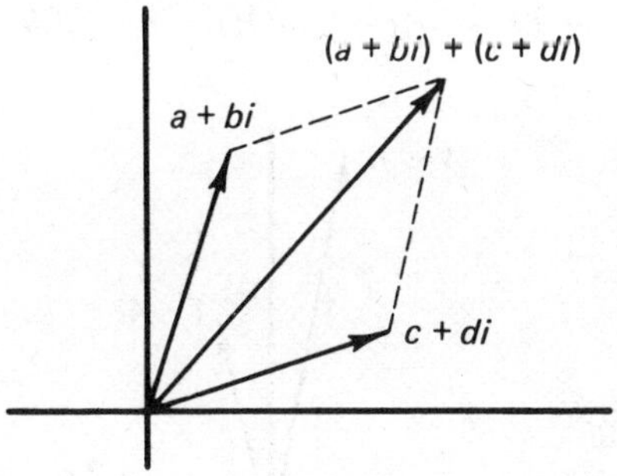

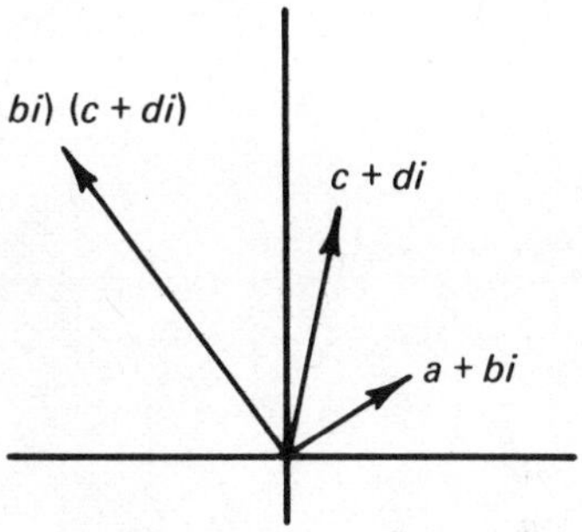

The sum of two numbers is the fourth vertex of a parallelogram with one vertex at the origin. The distance of the product from the origin is the product of the distances of the factors. The angle that the line from the origin to the product makes with the x-axis is the sum of the corresponding angles for the factors.

Example: The various sums and differences of the numbers $4 + i$ and $2 + 5i$ are shown below.

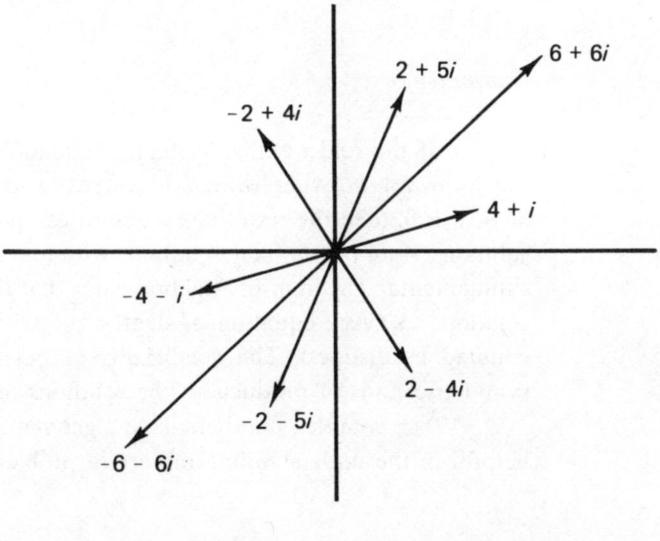

Example: The definitions of addition and multiplication simply formali the expected rules for working with $i = \sqrt{-1}$. (Notice that by definitic $i^2 = -1$.) The next diagram shows the results of multiplying some particul complex numbers.

$$(1 + 4i)(3 + i) = -1 + 7i$$
$$(-1 + 7i)(1 + 4i) = -29 + 3i$$
$$(-29 + 3i)(-1 + 7i) = 8 - 206i$$

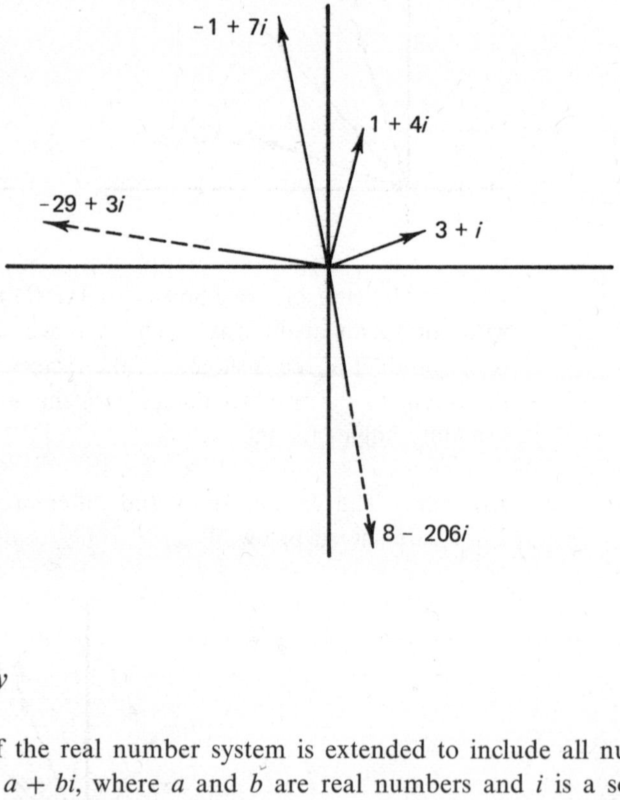

Summary

If the real number system is extended to include all numbers the form $a + bi$, where a and b are real numbers and i is a solution $x^2 + 1 = 0$, then the resulting system of complex numbers contains all th solutions of all polynomial equations with real or complex coefficients. Th Fundamental Theorem of Algebra states that there are n complex numb solutions to every equation of degree n (provided multiple solutions a counted as distinct). The coefficients of polynomials are expressible symmetric sums of products of the solutions of the polynomial equation

The complex numbers have a geometric interpretation that is ve helpful in the understanding of certain problems in the physical science

EXERCISES

1. Use Girard's expressions for the coefficients to find equations that have exactly the following collections of numbers as solutions. Check your answers by verifying that the numbers are solutions.

(a) 1, 2, 3 (c) 2, 2, i, $-i$

(b) $2 + i$, $2 - i$

2. What can be deduced from the theorems of this chapter about the solutions of

$$x^3 - 2x^2 - 5x + 6 = 0 \quad \text{and} \quad x^4 - 7x^3 + 11x^2 + 7x - 10 = 0.$$

Is it possible that either of these equations has solutions that are all integers?

DISCUSSION QUESTIONS

1. Investigate to determine which of the properties of the real numbers listed in Chapter 16 are also true for the complex numbers. In particular, which complex numbers are 0, 1, the additive inverses, and the multiplicative inverses?

2. Consider the set of cubic equations with real coefficients, $x^3 + ax^2 + bx + c = 0$.

(a) Show that geometric considerations imply that a cubic equation cannot have 0 or 2 real solutions.

(b) Show that if there is only one real solution, then the sum and product of the other two solutions are real numbers.

(c) Investigate the similarities and differences between members of a pair of complex solutions. Graph a few examples.

***3.** The absolute value of a real number may be interpreted geometrically as the distance from x to the origin. (By definition $|x| = x$ if $x \geqslant 0$ and $|x| = -x$ if $x < 0$.) Use the graphical representation of complex numbers to define an analogous concept of absolute value for complex numbers. Does your absolute value have properties similar to those of the absolute value for real numbers? For example, is it true that $|-c| = |c|$, $|c_1| \cdot |c_2| = |c_1c_2|$, and $|c_1 + c_2| \leqslant |c_1| + |c_2|$?

***4.** By the fundamental theorem the equation $x^n - 1 = 0$ should have exactly n solutions. What can you say about the nature of the solutions? Investigate by finding the roots explicitly for small values of n. [Note that $x^3 - 1 = (x - 1)(x^2 + x + 1)$.] It will help to consider the geometric representation of the solutions.

twenty

permutations

The formulas given by Girard for expressing the coefficients of polynomial $p(x)$ explicitly in terms of the solutions of the equation $p(x) =$ were studied by many mathematicians in the seventeenth and eighteent centuries. It was hoped that an investigation of these representations migh lead to a solution of the problem of representing the solutions of an polynomial equation explicitly in terms of the coefficients. The formula known in the quadratic, cubic, and quartic cases are complicated, but it wa hoped that similar formulas involving only the ordinary operations o arithmetic and the taking of roots might be found for the case of th polynomial of degree n by studying the formulas given by Girard.

Niels Henrik Abel (1802–1829) was the son of a Norwegian ministe While a student in high school he became interested in the problem o finding a general formula that could be used to solve all equations of degre five. At one point he was convinced and he convinced his teachers that h had done so. When he entered the university, he read the papers of Josep Louis Lagrange (1736–1813), who had worked for years on the problem and conjectured that it could not be done. Abel extended the ideas an methods of Lagrange to prove that indeed the search for such a formul was hopeless. In a paper published after some delay (1826), Abel prove that for equations of degree higher than four it is not true that the solution can always be expressed as an algebraic combination of the coefficient

Since Abel had no money at all, he was given a modest amour to cover the expenses of further study at the centers of mathematical learnin in Paris and Göttingen. He returned home without having been able t convince the eminent European mathematicians (such as Gauss) to read his work. He died at the age of 26, just before he became famous.

One young Frenchman who did read Abel's papers on the theory of equations was Evariste Galois (1811–1831). Galois was a prodigy who could not resist continually demonstrating his superiority to his teachers in mathematics. By the time they finally passed him on the oral examinations necessary for enrollment in a university, Galois had finished his major mathematical treatise. He had set himself the task of completing the work on equations started by Abel. He found the criteria that must be satisfied if the solutions to an equation are to be expressed as algebraic combinations of the coefficients.

Galois, like Abel, was unable to persuade prominent mathematicians to read his work. He became disgusted with the "establishment," joined the revolution in 1830, was provoked (by a government agent?) into a duel, and was killed before his twentieth birthday. Galois' paper was finally published in 1846. It provided the starting point for many major developments in algebra in the twentieth century.

The exact statements and the proofs of Abel's and Galois' theorems are difficult and we shall not discuss them. It should be noted, however, that there is a significant difference between the nature of these theorems and that of most of the theorems we have studied. Cardano and Ferrari exhibited formulas of a special kind for the solutions of cubic and quartic equations. Abel proved that such *formulas do not exist* for equations of degree five. (Compare the differences between proving a set of axioms inconsistent and proving it consistent.)

Sets of Permutations

An idea of the kind of mathematics used by Abel and Galois can be obtained by looking at the work of Lagrange. Lagrange was one of the first mathematicians to take full advantage of the power of algebraic symbolism. He invented symbols, used them to make clear the essential parts of problems, and then manipulated them to solve the problems.

Lagrange devoted his entire life to science and avoided worldly interests. This undoubtedly contributed to his longevity during a period when many aristocratic scholars were losing their heads to the French Revolution. Lagrange was (and is) considered to be the greatest mathematician of the eighteenth century and before his death he was awarded many high honors by the admiring Napoleon.

Lagrange studied the equations that give the coefficients in terı of the solutions for the equation $x^n + a_1x^{n-1} + \cdots + a_n = 0$.

$$
\begin{aligned}
-a_1 &= r_1 + r_2 + r_3 + \cdots + r_n \\
a_2 &= r_1r_2 + r_1r_3 + \cdots + r_1r_n + r_2r_3 + r_2r_4 + \cdots + r_{n-1}r_n \\
-a_3 &= r_1r_2r_3 + r_1r_2r_4 + \cdots + r_{n-2}r_{n-1}r_n \\
&\vdots \\
(-1)^n a_n &= r_1r_2 \ldots r_n
\end{aligned}
$$

One of the things he noticed about these equations is that each expressi for the coefficients is symmetric. If r_1 is written in place of r_2 and r_2 in pla of r_1, for example, in any of the expressions, then the equation remains tru The value of the expression is unchanged. This led Lagrange to a stu of the properties of permutations and of the effect they have on expressio of this kind.

DEFINITION | A **permutation** of the set $\{1, 2, \ldots, n\}$ is a one-to-one mapping of the s onto itself.

Since a permutation is a mapping, it is defined and described b specifying the image of every element of the set. There are a number special ways of denoting the permutations of a finite set. It is possible list the numbers and draw an arrow from each number to its image. Or of the more common notations, because it is easy to work with, is to wri the numbers in a row and then write a second row putting the image each number in the first row directly below the number. The most compa notation describes the permutation by making a list or lists in which ea number is followed immediately by its image. The last number in a list assumed to map into the first number of the list. For typographical an aesthetic reasons this last notation will be used in this text.

Example: Three of the permutations of the set $\{1, 2, 3, 4, 5\}$ are describe below with each of the notations. Parentheses are used to separate the lis or *cycles* of the method we shall use.

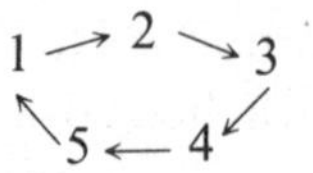

12345
23451

(12345)

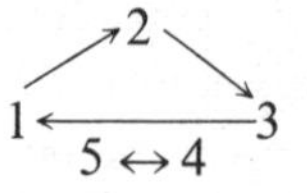

12345
23154

(123)(45)

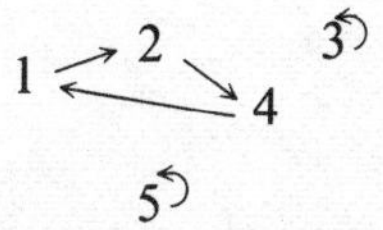

12345 (124)
24315

The special permutation that maps every element into itself is called the **identity permutation** and denoted I. The set of all permutations of the set $\{1, 2, \ldots, n\}$ is denoted S_n. It is easy to believe that there are $n! = n(n-1) \cdot \cdots 2 \cdot 1$ elements in S_n because there are n choices for the image of 1, and then $n - 1$ choices for the image of 2, and so on.

Examples: The arrays that follow show the two methods of denoting each of the members of S_2, S_3, and S_4. The columns on the left show the second row of the two-row notation. Immediately to the right is the cyclic notation for the same permutations.

$n = 2$

1	2	
1	2	I
2	1	(12)

$n = 3$

1	2	3	
1	2	3	I
1	3	2	(23)
2	1	3	(12)
2	3	1	(123)
3	1	2	(132)
3	2	1	(13)

$n = 4$

1	2	3	4	
1	2	3	4	I
1	2	4	3	(34)
1	3	2	4	(23)
1	3	4	2	(234)
1	4	2	3	(243)
1	4	3	2	(24)
2	1	3	4	(12)
2	1	4	3	(12)(34)
2	3	1	4	(123)
2	3	4	1	(1234)
2	4	1	3	(1243)
2	4	3	1	(124)

1	2	3	4	
3	1	2	4	(132)
3	1	4	2	(1342)
3	2	1	4	(13)
3	2	4	1	(134)
3	4	1	2	(13)(24)
3	4	2	1	(1324)
4	1	2	3	(1432)
4	1	3	2	(142)
4	2	1	3	(143)
4	2	3	1	(14)
4	3	1	2	(1423)
4	3	2	1	(14)(23)

Joseph Louis Lagrange. *Lagrange succeeded Euler as director of the Berlin Academy. He returned to France in 1786 and was president of the committee which devised the metric system of weights and measures during the French Revolution.*

Composition of Permutations

Lagrange noticed that each of the equations giving the coefficients of a polynomial in terms of its solutions is left invariant by *all* the possible permutations of the solutions. He then investigated other similar expressions to see what effect the permutations would have. He found that some very interesting patterns exist.

Example: Consider the expression $r_1 - r_2 - r_3$. There are six permutations of the numbers r_1, r_2, and r_3. Apply each of these permutations to the expression by replacing the numbers by their images. The permutations I and (23) leave the value of the expression unchanged. The permutations (12) and (123) change the expression to $r_2 - r_1 - r_3$. The remaining two permutations, (13) and (132), change the expression to $r_3 - r_1 - r_2$.

Example: Consider the expression below.

$$r_1^2 r_2 + r_2^2 r_3 + r_3^2 r_1$$

If the six permutations of S_3 are applied to the numbers in this expression, then it can be seen that three leave the expression unchanged and the other three all change it in the same way. The permutations I, (123), and (132) do not change the expression. The permutations (12), (13), and (32) change it to

$$r_2^2 r_1 + r_1^2 r_3 + r_3^2 r_1.$$

In this case the six permutations can be divided into two subsets of three members each according to the effect they have on the expression.

Example: Exactly eight of the 24 elements of S_4 leave the expression below invariant.

$$r_1 r_2 + r_3 r_4$$

Eight more change it to a second expression and the remaining eight change it to a third form. The permutations that leave this expression invariant are

$$I, (12), (34), (12)(34), (13)(24), (14)(23), (1324), (1423).$$

Example: The six permutations of S_4 that map 1 into itself (see p. 18 leave

$$r_1 - r_2 - r_3 - r_4$$

invariant. Clearly the six permutations that map 1 into 2, the six others th map 1 into 3, and the six that map 1 into 4 form subsets such that eac member of a subset has the same effect on this expression.

Lagrange made several conjectures from data like that provided b the examples and was able to prove some of his conjectures. The observatic that led eventually to the work of Abel and Galois is that not only can th permutations be divided into classes according to the result of applying the to a given expression, but those subsets of permutations that leave th expression unchanged have certain special properties. These properties hav to do with the idea of composite permutations.

DEFINITION | Let P_1 and P_2 be permutations of the set $\{1, 2, \ldots, n\}$. The **compositio of P_1 and P_2**, denoted $P_2 \cdot P_1$ or P_2P_1, is the permutation that is the resu of mapping first by P_1 and then by P_2.

Example: Consider the permutations (1234) and (12) from S_4. The compo sition of (12) and (1234), written (1234)(12), is (134). The computation ca be done by inspection or by writing down the following arrays.

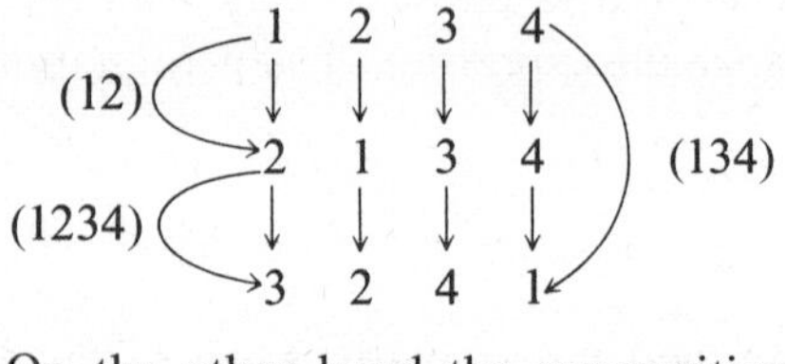

(1234)(12) = (134)

On the other hand the composition of (1234) and (12) is (234 Composition of permutations is not commutative.

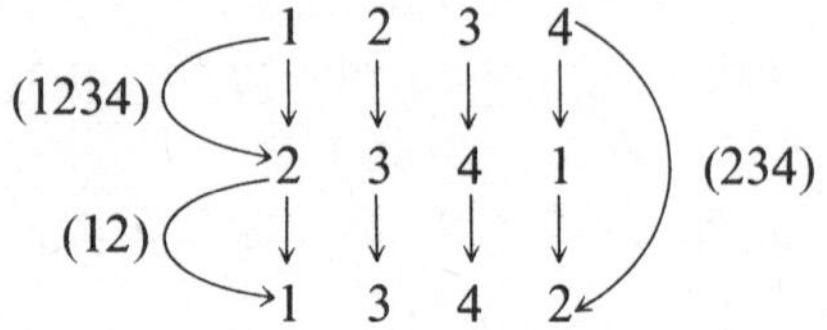

(12)(1234) = (234)

Example: It is often convenient to summarize the data about compositior in the form of tables similar to addition and multiplication tables. The tw tables following show some examples from S_3. The entry for (12)(13) (132) and (13)(12) = (123).

	I	(123)	(132)
I	I	(123)	(132)
(123)	(123)	(132)	I
(132)	(132)	I	(123)

	(12)	(13)	(23)
(12)	I	(132)	(123)
(13)	(123)	I	(132)
(23)	(132)	(123)	I

The composition of two permutations in S_n is another element of S_n. Furthermore, the composition of any two permutations from a subset that leaves some expression invariant is a permutation of that same subset. In other words, this kind of *subset* of S_n is *closed under composition.* A little thought about the common property of the permutations in the subset should convince you that the subset should indeed be closed.

The tables in the last example are the composition tables for the sets $\{I, (123), (132)\}$ and $\{(12), (13), (23)\}$. It should be clear that the first set is closed under composition and the second set is not.

Two additional characteristic properties of the sets of permutations that leave some expression invariant are described in the first discussion question at the end of this chapter.

Summary

Abel proved that a general formula that will express the solutions of all fifth-degree equations as algebraic combinations of the coefficients does not exist. Galois characterized the polynomial equations that have solutions expressible in such a form. The methods they used were based on the properties of special sets of permutations. The fact that there is a connection between the solvability of an equation and the properties of sets of permutations of the solutions had been suggested and investigated earlier by Lagrange.

In the next chapter permutations will be used to discuss a different kind of problem in algebra. Chapter 24 contains some further remarks about the properties of special sets of permutations, including an important theorem proved by Lagrange, and brief comments on the method used by Galois.

EXERCISES

1. Classify the permutations of S_3 by the effect they have on $r_1 + r_2 - r_3$. Construct a composition table for the set that leaves it unchanged. Verify that this subset is closed under composition.

2. Repeat Exercise 1 for S_4 and $r_1 + r_2 - r_3 - r_4$.

DISCUSSION QUESTIONS

1. Explain why the permutation I must always be in the set o permutations that leave an expression invariant. Explain why the invers of any permutation in this set is also in the set. (Notice that since a permuta tion is a transformation, it must have an inverse. See Chapter 10 for review Do the other subsets have this second property?

2. Investigate the following conjecture: "If any permutation repeated enough times, each number will be mapped to itself."

3. A *transposition* is a permutation in which each of the numbe is mapped to itself except for two that are mapped to each other. A permu tation is *even* if it can be written as the composition of an even numbe of transpositions and *odd* if it can be written as an odd number. Investiga each of the following statements.

(a) Every permutation can be written as a composition of tran positions.

(b) The inverse of a composition of transpositions is the compos tion in *reverse* order of the same transpositions.

*(c) Every permutation is either even or odd but not both.

(d) The set of even permutations is closed but the set of odd pe mutations is not closed.

*(e) Every permutation is expressible as a composition of odd pe mutations, but not every permutation is expressible as a compositio of even permutations.

wenty-one

sets of linear equations

The problem of solving simultaneously a set of n linear equations in n unknowns is one that occurs frequently in the practical applications of mathematics. The techniques for solving this kind of problem have been known for a long time but involve so many computations that it was impractical, until recently, to apply them to solve large sets of equations. The development of high-speed computing machines has made it possible to use these methods and stimulated interest in finding more efficient algorithms that can be used by machines to solve the problems.

The Method of Elimination

The problem in the case when $n = 2$ is to find the values of x and y, if any, that satisfy the two equations

$$\begin{aligned} a_1x + b_1y &= c_1 \\ a_2x + b_2y &= c_2. \end{aligned}$$

The method that is taught in schools today has been used since the time of the Babylonians. The procedure is to eliminate one of the unknowns by combining the equations. This results in a single linear equation in one unknown, which can easily be solved. The second unknown can then be found by substitution.

Example: If the system of equations is

$$\begin{aligned} 3x + 4y &= 1 \\ 2x + 5y &= -4 \end{aligned}$$

then the result of multiplying both sides of the equations by 5 and respectively, is

$$15x + 20y = 5$$
$$8x + 20y = -16.$$

If the second equation is subtracted from the first, the result is $7x = 2$ Thus, $x = 3$. Then $9 + 4y = 1$ and $y = -2$. Of course the value of y cou also be found by using an appropriate elimination procedure on the origin problem.

If this same procedure is used on the general problem, then th steps are

$$b_2a_1x + b_2b_1y = c_1b_2$$
$$b_1a_2x + b_1b_2y = c_2b_1$$
$$(a_1b_2 - a_2b_1)x = c_1b_2 - c_2b_1$$
$$x = \frac{c_1b_2 - c_2b_1}{a_1b_2 - a_2b_1}$$

If the corresponding elimination is used to solve for y, the result is

$$y = \frac{a_1c_2 - a_2c_1}{a_1b_2 - a_2b_1}.$$

It would be impractical to go through all these computations to solve an particular problem, but the method has provided a formula that can be use to find exactly one solution to a set of two equations in two unknowns *excep* when $a_1b_2 - a_2b_1 = 0$.

Example: The solution of the system

$$x + 3y = 7$$
$$4x - y = 2$$

is given by the formulas as

$$x = \frac{7(-1) - 2 \cdot 3}{1(-1) - 4 \cdot 3} = \frac{-13}{-13} = 1 \qquad y = \frac{1 \cdot 2 - 4 \cdot 7}{1(-1) - 4 \cdot 3} = \frac{-26}{-13} = 2.$$

The geometric interpretation of a system of two equations in tw unknowns will give some insight into the nature of the solutions. The grap of each equation is a straight line. Usually two lines intersect at exactly on point. The coordinates of that point are the solution. There are two other possibilities.

If the lines are different, but parallel, then they do not intersect and the system has no solutions. If the two equations represent the same line, then every point on that line is a solution of the system. From the computations above it follows that $a_1b_2 - a_2b_1$ must be zero in the last two cases. This is true because the equation $a_1b_2 - a_2b_1 = 0$ is equivalent to $-a_1/b_1 = -a_2/b_2$, which states that the slopes of the lines are the same.

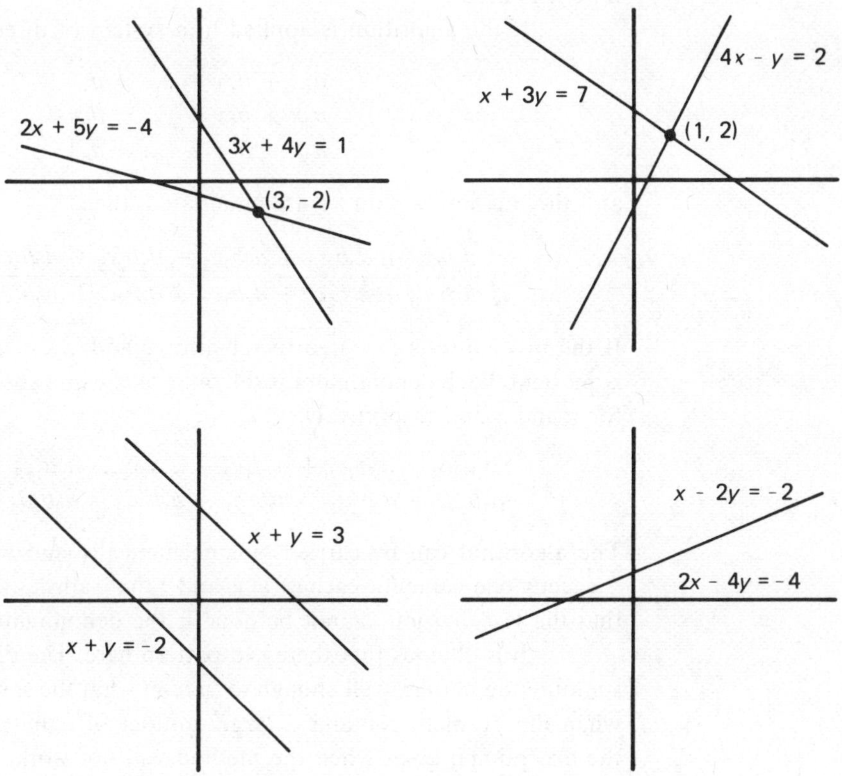

It is possible to generalize this method for two equations and obtain an algorithm that can be used to solve systems having more equations. Although we shall have no use for the ability to solve larger systems, the spirit of mathematics requires that we be certain that such systems can be solved and that it can be done efficiently.

No matter what the value of n, one can combine the first equatio of a set with each of the others to eliminate the first unknown. The resul is a system with one less equation and each equation contains one les unknown. The equations of this new system can be combined in the sam way to reduce the problem again. Eventually, there will be just two equation left and the formulas for the case $n = 2$ can be applied.

This algorithm obviously requires many computations. It is desirabl to have formulas similar to those in the case when $n = 2$ so that it will no be necessary to go through the details of the elimination algorithm ever time it is used.

If the algorithm is applied to a system of three equations,

$$\begin{aligned} a_1x + b_1y + c_1z &= d_1 \\ a_2x + b_2y + c_2z &= d_2 \\ a_3x + b_3y + c_3z &= d_3 \end{aligned}$$

and the unknowns y and z are eliminated, then

$$x = \frac{d_1b_2c_3 + d_2b_3c_1 + d_3b_1c_2 - d_1b_3c_2 - d_2b_1c_3 - d_3b_2c_1}{a_1b_2c_3 + a_2b_3c_1 + a_3b_1c_2 - a_1b_3c_2 - a_2b_1c_3 - a_3b_2c_1}.$$

If the procedure is repeated to solve for y and z, the answer each time i a quotient. Each denominator is the same as the one above. The numerator for y and z are, respectively,

$$\begin{gathered} a_1d_2c_3 + a_2d_3c_1 + a_3d_1c_2 - a_1d_3c_2 - a_2d_1c_3 - a_3d_2c_1 \\ a_1b_2d_3 + a_2b_3d_1 + a_3b_1d_2 - a_1b_3d_2 - a_2b_1d_3 - a_3b_2d_1 \end{gathered}$$

The algorithm can be carried out mechanically and will show that ther is exactly one value for each of x, y, and z that satisfies the equations *excep* that the last division cannot be done if the denominator above is zero.

It is obvious that there is a pattern here. The difficulty is in under standing the pattern well enough to predict what the solutions will look lik when the problem contains a large number of equations and to identif the exceptional cases when the method will not work.

Cramer's Rule

By the end of the seventeenth century the notation for algebrai equations had been developed in Europe to the point that it was possibl to notice that there is a pattern in the form of the solutions of systems o linear equations. The Scottish mathematician Colin Maclaurin (1698–1746) included a description of the solutions in terms of quotients in his book on algebra. This result is now known as *Cramer's Rule* in honor of the Swiss geometer Gabriel Cramer (1704–1752) who published his version a few years later.

Cramer's main interest was in the properties of geometric figures. He used the techniques of analytic geometry to classify curves according to their algebraic equations. *Cramer's Rule* appears in an appendix (of algebraic facts) to a book by Cramer on geometry. The notation he used is almost exactly the same as that used in the preceding section for the cases $n = 2$ and $n = 3$. Cramer included a description of how these quotients can be constructed from the coefficients in the general case.

Before we introduce some notation and state Cramer's Rule as a formal theorem, it will be helpful to consider in detail what the rule should say. The following statements are therefore intended not only to completely describe the formulas we have found when $n = 2$ and $n = 3$, but also to provide the information necessary to write down the formulas for solutions of larger systems.

Each component of a solution is a quotient. The common denominator that occurs is the most important quantity because it determines whether or not the method will work. The different numerators can be obtained from the denominator by making appropriate substitutions. More precisely, the numerator of the solution for a given unknown can be obtained by replacing the coefficients of that unknown by the constants of the equations. Each coefficient is replaced by the constant term from the same equation as the coefficient.

The denominator consists of a sum and difference of products. Each product consists of one coefficient from each unknown, chosen so that each equation is represented. If the coefficients are written in the order in which the unknowns occur in the equations, then the subscripts, which indicate the equation in which the coefficients occur, form every possible permutation. Half of the products are subtracted from the sum of the other half. Whether or not a product is to be added or subtracted depends on the permutation. The term is added if the permutation can be done with an even number (including zero) of transpositions and subtracted if the number of transpositions is odd.

The preceding paragraphs contain a complete description of the rule for computing the unique solution to a set of n linear equations in n unknowns when such a solution exists. In order to obtain a precise formula and make manipulation of the quantities involved possible, it is necessar to make some definitions and introduce some notation. (This process i supposed to make the description clearer and computations easier. Th ability to read mathematical notation is, however, an acquired skill. Th following notation is included here to show how it can and would be done even though it is doubtful that you will find it of significant value. You wil not be able to appreciate it without some careful study.)

DEFINITION | A **matrix** is a rectangular array of numbers. The entries in a matrix wil be described by subscripts showing the row and column in which the entr occurs. The square matrix below with n rows and columns will be denote $[a_{ij}]_n$.

$$\begin{bmatrix} a_{11} & a_{12} & \cdots & a_{1n} \\ a_{21} & a_{22} & \cdots & a_{2n} \\ & & \vdots & \\ a_{n1} & a_{n2} & \cdots & a_{nn} \end{bmatrix}$$

DEFINITION | If $A = [a_{ij}]_n$ is a square matrix, then the **determinant of A,** denoted $|A|$, i the sum of terms of the form

$$(-1)^k a_{1-} \cdot a_{2-} \cdot a_{3-} \cdots a_{n-}$$

There are $n!$ terms. Each term has a different permutation of the secon subscripts. The exponent k is the number 1 or 2 according to whether th permutation is an odd or even number of transpositions.

We can now use the definitions and notation to concisely restat the results we have found about the solutions of systems of equations.

THEOREM (Cramer's Rule) | The system of equations

$$\begin{aligned} a_{11}x_1 + a_{12}x_2 + \cdots + a_{1n}x_n &= b_1 \\ a_{21}x_1 + a_{22}x_2 + \cdots + a_{2n}x_n &= b_2 \\ &\vdots \\ a_{n1}x_1 + a_{n2}x_2 + \cdots + a_{nn}x_n &= b_n \end{aligned}$$

has a unique solution if and only if the determinant of the matrix $A = [a_{ij}]$ is not zero. Moreover, if A_j is the matrix obtained by replacing the jth colum of A with a column containing the numbers $b_1, b_2, \ldots, b_n$, then the compo nents of the solution are

$$x_1 = \frac{|A_1|}{|A|}, x_2 = \frac{|A_2|}{|A|}, \ldots, x_n = \frac{|A_n|}{|A|}.$$

Example: Cramer's Rule states that since the determinant of the matrix of coefficients is not zero, there is exactly one solution of the system below.

$$\begin{aligned} 2x_1 - 7x_2 &= 11 \\ 3x_1 + 4x_2 &= 8 \end{aligned} \qquad \begin{vmatrix} 2 & -7 \\ 3 & 4 \end{vmatrix} = 29$$

The rule states further that the solution is

$$x_1 = \frac{\begin{vmatrix} 11 & -7 \\ 8 & 4 \end{vmatrix}}{\begin{vmatrix} 2 & -7 \\ 3 & 4 \end{vmatrix}} = \frac{100}{29} \qquad x_2 = \frac{\begin{vmatrix} 2 & 11 \\ 3 & 8 \end{vmatrix}}{\begin{vmatrix} 2 & -7 \\ 3 & 4 \end{vmatrix}} = \frac{-17}{29}.$$

Example: Cramer's Rule shows that because the determinant is zero, the system

$$\begin{aligned} 2x - 4y &= 3 \\ 4x - 8y &= 6 \end{aligned} \qquad \begin{vmatrix} 2 & -4 \\ 4 & -8 \end{vmatrix} = 0$$

does not have a unique solution. These two equations define the same line and there are an infinite number of pairs (x, y) that satisfy the system.

The computations needed to find the values of determinants are very cumbersome. As a result, Cramer's Rule is seldom used to actually find the solutions of a system of equations. Instead, the properties of matrices and determinants have been studied extensively and a number of techniques have evolved that make it possible to solve systems of equations by manipulating matrices. The subject matter of this and the next two chapters are part of the subdivision of mathematics called *linear algebra.*

Summary

Elimination methods have been used for millennia to solve sets of linear equations. Eventually the improvements made in the notation used to state and work such problems enabled mathematicians to see that the solutions could be written in terms of the coefficients according to a general pattern that holds no matter how many equations are involved. Cramer's

Rule is a theorem that states the exact condition under which a unique solution exists and gives a complete description of that solution. The introduction of the concepts of determinant and matrix makes it possible to state Cramer's Rule concisely.

EXERCISES

1. Use Cramer's Rule to determine whether or not the following systems have unique solutions. If so, then use the rule to find the solution. If not, give a geometric interpretation of the system.

$$\begin{aligned} 7x - 2y &= 3 \\ -4x + y &= 8 \end{aligned} \qquad \begin{aligned} 2x - y &= 3 \\ -6x + 3y &= -9 \end{aligned}$$

$$\begin{aligned} 3x + 2y &= 4 \\ 6x + 4y &= 10 \end{aligned} \qquad \begin{aligned} x + y &= 1 \\ x - y &= 1 \end{aligned}$$

2. Check your understanding of Cramer's Rule further by making up a few problems and solving them. To obtain simple problems, select the answers in advance. In the case when $n = 2$, first choose some coefficients, such as

$$\begin{aligned} &3x + 2y \\ &5x - 7y \end{aligned}$$

and then choose some solutions, such as $x = 1$ and $y = 3$; then you will need 9 and -16 as the constants in the equations. You now have a problem for which you know the answer.

DISCUSSION QUESTIONS

1. Is there a geometric interpretation of a system of three equations in three unknowns? Investigate what the three-dimensional graph of an equation $ax + by + cz = d$ looks like by checking several trivial examples like $z = 0$. Make as many conjectures as you can about the nature of the solutions of such systems.

2. Investigate the connection between the *size* (absolute value) of the number

$$\frac{1}{2}\begin{vmatrix} x_1 & x_2 \\ y_1 & y_2 \end{vmatrix}$$

and the triangle that has one vertex at the origin and the other two vertices at the points (x_1, y_1) and (x_2, y_2). Begin by considering special cases such as right triangles that have one side on an axis. The diagram below may help you to justify your interpretation for the general case.

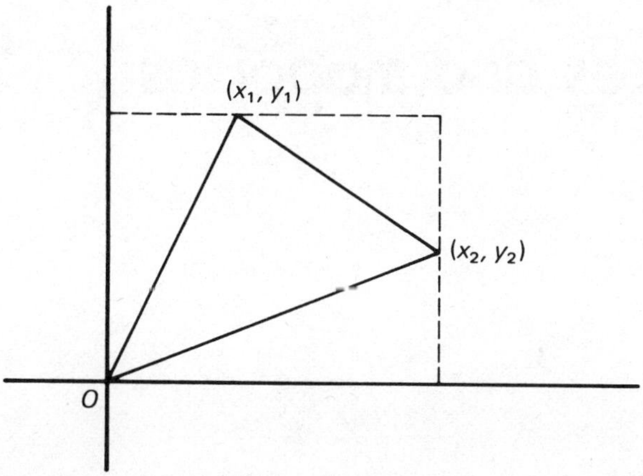

Find a similar determinant that can be used for triangles with three arbitrary vertices (x_1, y_1), (x_2, y_2), and (x_3, y_3). (*Hint*. Use a translation to reduce the problem to the one just solved. Refer to Chapter 10 if necessary.)

twenty-two

matrices and mappings

One of the main reasons that mathematicians of the nineteenth century were interested in the solutions of systems of equations is that such systems occur in the study of geometric transformations. The transformations themselves were of interest because of their usefulness in finding and simplifying the equations for the motion of planets and other objects in space. It was noted in Chapter 10 that if the plane is given Cartesian coordinates, then the methods of analytic geometry can be used to study transformations. In this chapter we shall consider those transformations for which the definitions, in the notation of analytic geometry, are sets of linear equations.

Matrix Representation of Mappings

If the mapping T, of the plane into itself, is defined by $T(x, y) = (x', y')$ where

$$x' = a_1x + b_1y$$
$$y' = a_2x + b_2y,$$

then it is easy to find the image (x', y') of any point (x, y) by substitution in the equations. By using the techniques for solving linear equations it is also possible to find the point (x, y) that a transformation maps into a given image (x', y'). (The words *mapping* and *transformation* are being used carefully here. Not all mappings of the form above are transformations. See the definition on p. 92.)

Example: For the transformation $T(x, y) = (3x + 2y, 2x + y)$ the images $T(1, 0) = (3, 2)$ and $T(11, -6) = (21, 16)$ are obvious. The point that has image (6, 7) is the solution of the system

$$\begin{aligned} 3x + 2y &= 6 \\ 2x + y &= 7. \end{aligned}$$

The solution is $(8, -9)$. It is easily checked that $T(8, -9) = (6, 7)$.

Arthur Cayley (1821–1895) was one of the foremost algebraists of the nineteenth century. He was the first to develop the analytic geometry for higher dimensions. His extensive algebraic work with transformations had a direct influence on Klein's decision to define geometry in terms of these functions. After graduating from Cambridge with high honors in mathematics, Cayley studied and practiced law until he was forty-two years old rather than submit to the religious formalities required of those who taught at Cambridge. During this period before the regulations were changed and he could accept a professorship, Cayley devoted every spare moment to mathematical research. In 1858 he published a paper entitled *Memoir on the Theory of Matrices* that revolutionized the study of transformations. As an indication of the kind of applied problems with which Cayley was working, we note that he published that same year a paper on the simplification of the equation for elliptical orbits.

Cayley worked with mappings, such as those we have seen above for the plane, in which the coordinates of the image are linear combinations of the coordinates of the point being mapped. He recognized that these mappings are completely determined by the choice of coefficients. In particular, Cayley noted that the essential facts about a mapping of the plane of the form $T(x, y) = (a_1x + b_1y, a_2x + b_2y)$ can be displayed in the matrix

$$\begin{bmatrix} a_1 & b_1 \\ a_2 & b_2 \end{bmatrix}.$$

Examples: The transformation $T_1(x, y) = (2x, y)$ that "stretches" by a factor of 2 in the horizontal direction has the matrix

$$T_1 : \begin{bmatrix} 2 & 0 \\ 0 & 1 \end{bmatrix}$$

The more complicated transformation $T_2(x, y) = (x + 2y, 3x + 4y)$ has the matrix

$$T_2 : \begin{bmatrix} 1 & 2 \\ 3 & 4 \end{bmatrix}$$

Arthur Cayley. *Cayley was one of the first to study matrices. His discovery of the laws for their manipulation marks him as one of the earliest investigators of abstract algebra.*

The "projection onto the x-axis," $T_3(x, y) = (x, 0)$, has the matrix

$$T_3 : \begin{bmatrix} 1 & 0 \\ 0 & 0 \end{bmatrix}$$

The identity transformation that maps every point onto itself, $T(x, y) = (x, y)$, has the matrix called the **identity matrix** and denoted I.

$$I = \begin{bmatrix} 1 & 0 \\ 0 & 1 \end{bmatrix}$$

Cayley invented the notation for matrices and used it to represent transformations. He investigated and discovered the properties of transformations by studying the matrices. The idea of using the array of coefficients to represent a system of linear conditions is a rather natural one. Cayley's concept of the matrix as a *single entity* that can be manipulated algebraically is more subtle.

Multiplication of Matrices

We have seen in our brief look at permutations and in our earlier discussion of geometric transformations that it is often convenient to combine mappings by doing them one after another. The formal definition of such combinations is the same as the one made in the special case of permutations.

DEFINITION | Let T_1 and T_2 be mappings on the set R. The **composition of T_1 and T_2**, denoted $T_2 \cdot T_1$ or T_2T_1, is the mapping obtained by applying first T_1 and then T_2.

Example: Consider the mappings T_1 and T_2 with matrices below.

$$T_1 : \begin{bmatrix} 1 & 2 \\ 3 & 4 \end{bmatrix} \qquad T_2 : \begin{bmatrix} 2 & 0 \\ 0 & 1 \end{bmatrix}$$

The images of the points (1, 0) and (5, 6) under the mapping T_2T_1 can be computed by repeated substitutions.

$$(T_2T_1)(1, 0) = T_2(T_1(1, 0)) = T_2(1, 3) = (2, 3)$$
$$(T_2T_1)(5, 6) = T_2(T_1(5, 6)) = T_2(17, 39) = (34, 39).$$

On the other hand the image of (1, 0) under T_1T_2 is

$$(T_1T_2)(1, 0) = T_1(T_2(1, 0)) = T_1(2, 0) = (2, 6).$$

The following diagram shows the effect of these mappings on some other particular points.

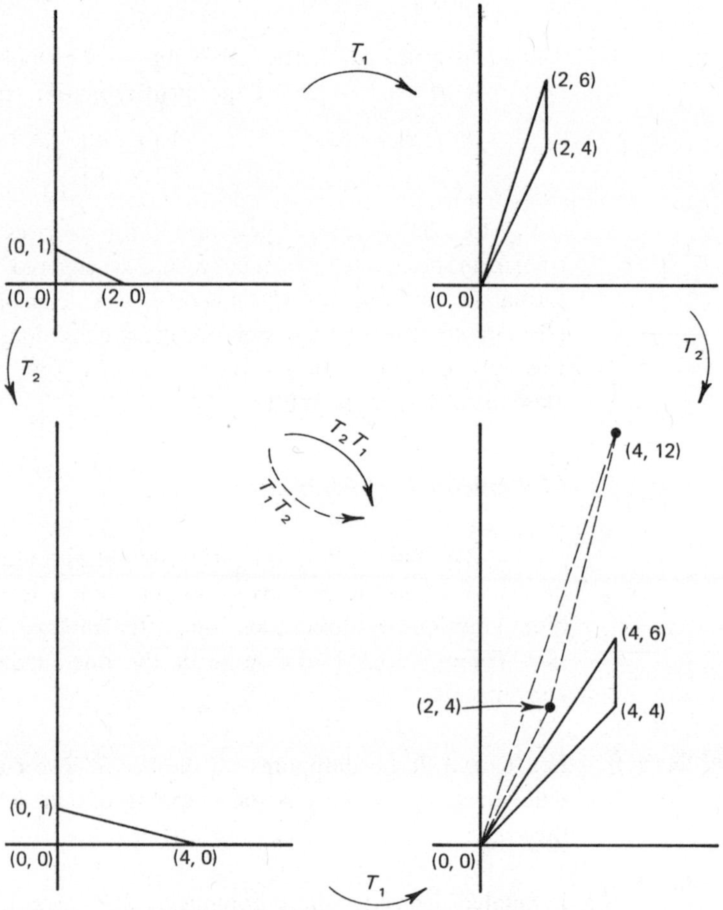

In general, if T_1 and T_2 are mappings that have the matrices

$$T_1 : \begin{bmatrix} a_1 & b_1 \\ a_2 & b_2 \end{bmatrix} \qquad T_2 : \begin{bmatrix} A_1 & B_1 \\ A_2 & B_2 \end{bmatrix},$$

then a little computation shows that

$$(T_2T_1)(x, y) = T_2(a_1x + b_1y,\ a_2x + b_2y)$$

and the right-hand side is equal to

$$(A_1a_1x + B_1a_2x + A_1b_1y + B_1b_2y,\ A_2a_1x + B_2a_2x + A_2b_1y + B_2b_2y).$$

This shows that the composition T_2T_1 is also a linear mapping that can be described by a two-dimensional matrix. It is then natural to define the **product of two matrices** to conform to this result.

$$\begin{bmatrix} A_1 & B_1 \\ A_2 & B_2 \end{bmatrix} \cdot \begin{bmatrix} a_1 & b_1 \\ a_2 & b_2 \end{bmatrix} = \begin{bmatrix} A_1a_1 + B_1a_2 & A_1b_1 + B_1b_2 \\ A_2a_1 + B_2a_2 & A_2b_1 + B_2b_2 \end{bmatrix}$$

Example: Using T_1 and T_2 as in the preceding example, we have

$$T_2T_1 : \begin{bmatrix} 2 & 0 \\ 0 & 1 \end{bmatrix} \cdot \begin{bmatrix} 1 & 2 \\ 3 & 4 \end{bmatrix} = \begin{bmatrix} 2 & 4 \\ 3 & 4 \end{bmatrix}$$

$$T_1T_2 : \begin{bmatrix} 1 & 2 \\ 3 & 4 \end{bmatrix} \cdot \begin{bmatrix} 2 & 0 \\ 0 & 1 \end{bmatrix} = \begin{bmatrix} 2 & 2 \\ 6 & 4 \end{bmatrix}$$

The facts that $(T_2T_1)(1, 0) = (2, 3)$ and $(T_1T_2)(1, 0) = (2, 6)$ can now be seen by inspection.

Example: If A is the matrix for the reflection in the x-axis, then we have

$$A = \begin{bmatrix} 1 & 0 \\ 0 & -1 \end{bmatrix} \quad A^2 = \begin{bmatrix} 1 & 0 \\ 0 & -1 \end{bmatrix} \cdot \begin{bmatrix} 1 & 0 \\ 0 & -1 \end{bmatrix} = \begin{bmatrix} 1 & 0 \\ 0 & 1 \end{bmatrix} = I$$

$$A^3 = A^2 \cdot A = \begin{bmatrix} 1 & 0 \\ 0 & 1 \end{bmatrix} \cdot \begin{bmatrix} 1 & 0 \\ 0 & -1 \end{bmatrix} = \begin{bmatrix} 1 & 0 \\ 0 & -1 \end{bmatrix} = A.$$

The geometric interpretation of the values of A^2 and A^3 is worth noticing.

The composition of the identity transformation, in either order, with some other transformation T is just T. This implies and is implied by the fact that the identity matrix has the property that for every matrix A,

$$A \cdot I = I \cdot A = A.$$

The definition of a transformation as a one-to-one mapping ensures that every transformation T must have an inverse T^{-1} such that the composition $T^{-1}T$ is the identity transformation.

The mapping with matrix

$$\begin{bmatrix} a_1 & b_1 \\ a_2 & b_2 \end{bmatrix}$$

is a transformation if and only if the set of equations

$$\begin{aligned} a_1x + b_1y &= x' \\ a_2x + b_2y &= y' \end{aligned}$$

has a unique solution for every choice of x' and y'. The unique solution for a particular choice of x', y' provides the coordinates of the *one* point that maps to (x', y'). Cramer's Rule states that this occurs if and only if the determinant of the matrix of the mapping is not zero.

It is far from obvious that the inverse of a transformation that is representable by a matrix must also be representable by a matrix. If a matrix exists for T^{-1}, then it must be of the form

$$\begin{bmatrix} r & s \\ t & u \end{bmatrix}$$

where r, s, t, and u are numbers such that

$$\begin{bmatrix} r & s \\ t & u \end{bmatrix} \cdot \begin{bmatrix} a_1 & b_1 \\ a_2 & b_2 \end{bmatrix} = \begin{bmatrix} 1 & 0 \\ 0 & 1 \end{bmatrix}.$$

If the definition of multiplication is applied and the resulting entries equated with those in the matrix I, then

$$\begin{aligned} a_1r + a_2s &= 1 \qquad & b_1r + b_2s &= 0 \\ a_1t + a_2u &= 0 \qquad & b_1t + b_2u &= 1. \end{aligned}$$

If two of these equations are solved for r and s and the other two for t and u, then the matrix for the inverse transformation is found to be

$$\begin{bmatrix} \dfrac{b_2}{a_1b_2 - a_2b_1} & \dfrac{-b_1}{a_1b_2 - a_2b_1} \\ \dfrac{-a_2}{a_1b_2 - a_2b_1} & \dfrac{a_1}{a_1b_2 - a_2b_1} \end{bmatrix}$$

Since T is a transformation, the determinant $a_1b_2 - a_2b_1$ is not zero and the inverse transformation has this matrix.

We have shown that a matrix has a **multiplicative inverse** if and only if its determinant is not zero. In other words, a linear mapping has an inverse (is a transformation) if and only if its matrix has an inverse. The matrix of the inverse transformation is the inverse of the matrix of the transformation.

Example: Consider the mapping U that has the matrix below. U is a transformation because the determinant of the matrix is 5. The matrix U^{-1} can be found by using the array above.

$$U : \begin{bmatrix} 2 & -1 \\ 3 & 1 \end{bmatrix} \qquad U^{-1} : \begin{bmatrix} \frac{1}{5} & \frac{1}{5} \\ -\frac{3}{5} & \frac{2}{5} \end{bmatrix}$$

The diagram below shows the effects of U and U^{-1} on a particular triangle and illustrates the fact that $U \cdot U^{-1} = U^{-1} \cdot U = I$.

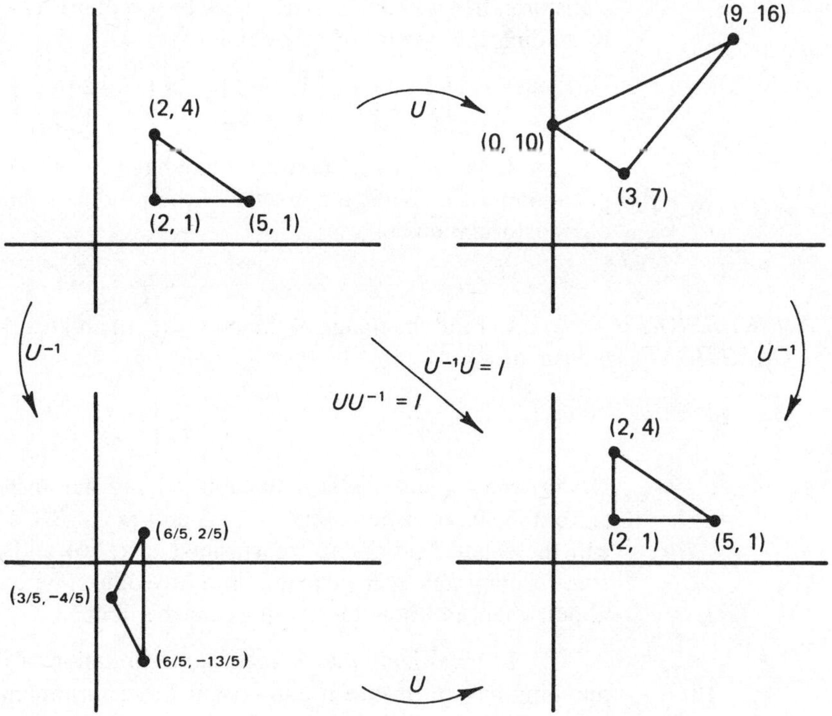

Summary

The transformations that are defined by linear equations are called *linear transformations*. There is a one-to-one correspondence between the set of matrices that have inverses and the set of linear transformations. The properties of a linear transformation are completely determined by the entries in the matrix. Any statement about matrices can be translated into a statement about linear transformations, and conversely.

An algebraic structure has been developed that can be used to solve problems involving geometric transformations. In particular, the multiplication of matrices corresponds exactly to composition of transformations. The next chapter is devoted to the investigation, using matrix methods, of some Euclidean transformations of the plane.

EXERCISES

1. Consider the mappings of the Euclidean plane that have the matrices below. Verify that three of them are transformations. Find the inverse of each transformation, and check your answer by composing each transformation with its inverse. Check the geometric effect of these mappings by finding the images of a few points.

$$\begin{bmatrix} 2 & 0 \\ 0 & 2 \end{bmatrix} \quad \begin{bmatrix} 1 & 0 \\ 0 & -1 \end{bmatrix} \quad \begin{bmatrix} 1 & -2 \\ 2 & 1 \end{bmatrix} \quad \begin{bmatrix} 1 & 2 \\ 2 & 4 \end{bmatrix}$$

2. Is the set of matrices that have inverses closed under multiplication? If so, how is the inverse of the product formed? (Think in terms of transformations.)

DISCUSSION QUESTIONS

1. Find the image of the points (1, 0) and (0, 1) under the mapping with matrix

$$\begin{bmatrix} a_1 & b_1 \\ a_2 & b_2 \end{bmatrix}.$$

Under what conditions is it possible to find a transformation that maps (1, 0) and (0, 1) to the points (x_1, y_1) and (x_2, y_2)? Under what conditions can these latter points be transformed to (1, 0) and (0, 1)? What about transforming any pair of points into any other? Be as specific as possible about when and how these things can be done.

2. Investigate the geometric interpretation of the fact that a linear mapping has an inverse if and only if the determinant of its matrix is not zero. (*Hint.* Recall Discussion Question 2 of Chapter 21.)

twenty-three

properties of linear transformations

It was shown in the preceding chapter that there is a one-to-one correspondence between linear transformations and invertible matrices. Cayley studied the multiplicative properties of matrices because he was interested in the geometric properties of the associated transformations. Both Cayley and Cramer used transformations to map geometric curves into positions in the plane where the curve equations would be least complicated. By this means they were able to isolate the characteristic features of the equations for different types of curves.

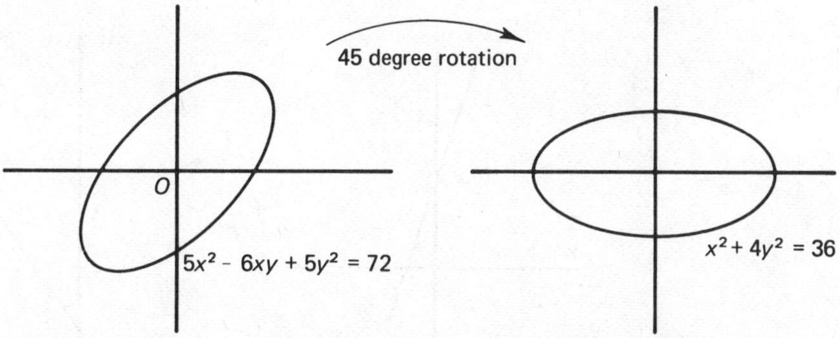

In this chapter we shall demonstrate the use of linear algebra by discussing briefly the linear transformations that are Euclidean in the sense of Klein. The objective of the discussion is a complete description of the matrices of those transformations that are similarities and isometries.

Images of Straight Lines

We have assumed in our casual investigation of the effects of linear transformations that straight lines are always mapped into straight lines. We shall confirm this now. A few specific examples should make it easier to appreciate the proof.

Examples: Consider the transformation $T(x, y) = (3x + y, 5x + 2y)$.

$$T : \begin{bmatrix} 3 & 1 \\ 5 & 2 \end{bmatrix} \qquad T^{-1} : \begin{bmatrix} 2 & -1 \\ -5 & 3 \end{bmatrix}$$

The set of points (x, y) that map onto the x-axis is the set for which the second coordinate of the image is zero. This is the line of points for which $5x + 2y = 0$. Similarly the y-axis is the image of the line $3x + y = 0$. Furthermore, the lines $x = 4$ and $y = -7$ are the images of the lines $3x + y = 4$ and $5x + 2y = -7$.

The x-axis ($y = 0$) is mapped by T to the curve that is mapped by T^{-1} onto the x-axis. An inspection of T^{-1} shows that it maps the line $-5x + 3y = 0$ onto $y = 0$. Therefore, $-5x + 3y = 0$ is the image of the x-axis under T. Similarly T maps the y-axis to the line $2x - y = 0$.

The diagrams below illustrate the effect of T on other lines.

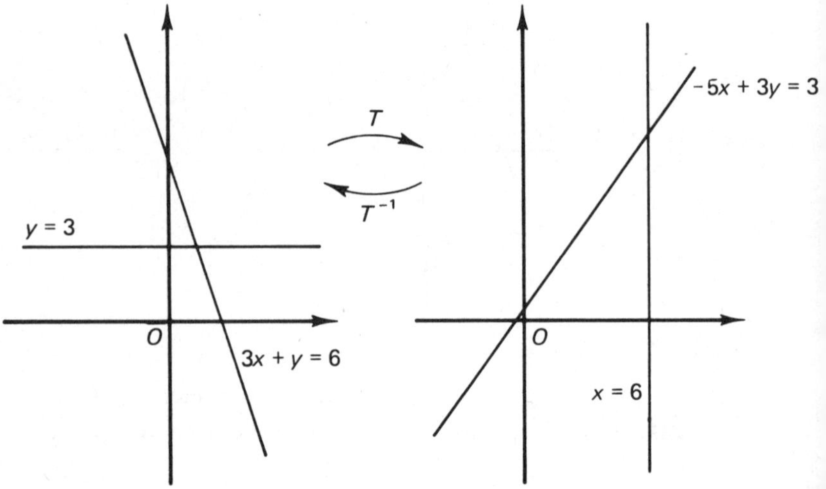

We shall now consider the general situation. Let U be the linear transformation

$$U(x, y) = (a_1x + b_1y, a_2x + b_2y) \qquad U: \begin{bmatrix} a_1 & b_1 \\ a_2 & b_2 \end{bmatrix}.$$

If the image of the point (r, s) lies on the line $Ax + By = C$, then

$$A(a_1r + b_1s) + B(a_2r + b_2s) = C$$

or

$$(Aa_1 + Ba_2)r + (Ab_1 + Bb_2)s = C.$$

Thus (r, s) lies on the line $(Aa_1 + Ba_2)x + (Ab_1 + Bb_2)y = C$.

In other words, those things that U maps onto straight lines are themselves straight lines. Since the inverse of U has this same property, it must be true that U maps straight lines—and only straight lines—onto straight lines.

Since the origin is a fixed point (maps onto itself), the images of the coordinate axes can be determined by finding the images of the points (1, 0) and (0, 1). If the transformation has matrix

$$\begin{bmatrix} a_1 & b_1 \\ a_2 & b_2 \end{bmatrix},$$

then the image of (1, 0) is (a_1, a_2) and the image of (0, 1) is (b_1, b_2). This information makes it possible to very easily describe some transformations.

The transformation that rotates the axes one quarter turn counterclockwise has matrix

$$R_1 = \begin{bmatrix} 0 & -1 \\ 1 & 0 \end{bmatrix}$$

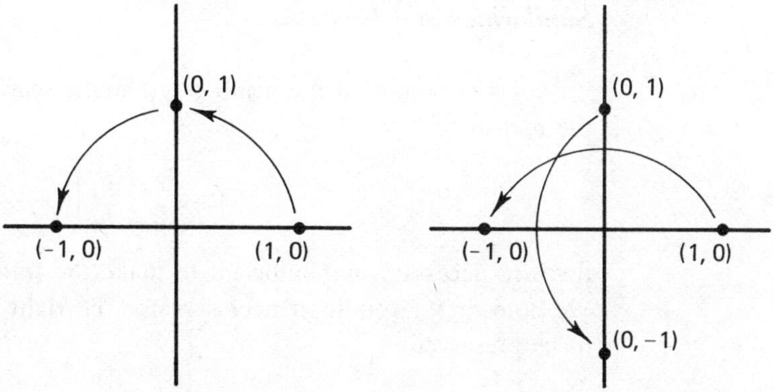

$R_1(x, y) = (-y, x)$ $\qquad$ $R_2(x, y) = (-x, -y)$

Also, the rotations through one-half and three-quarters of a circle are

$$R_2 = \begin{bmatrix} -1 & 0 \\ 0 & -1 \end{bmatrix} = R_1^2 \qquad R_3 = \begin{bmatrix} 0 & 1 \\ -1 & 0 \end{bmatrix} = R_1^3.$$

In addition, the reflections in the coordinate axes have matrices

$$F_1 = \begin{bmatrix} 1 & 0 \\ 0 & -1 \end{bmatrix} \qquad F_2 = \begin{bmatrix} -1 & 0 \\ 0 & 1 \end{bmatrix}.$$

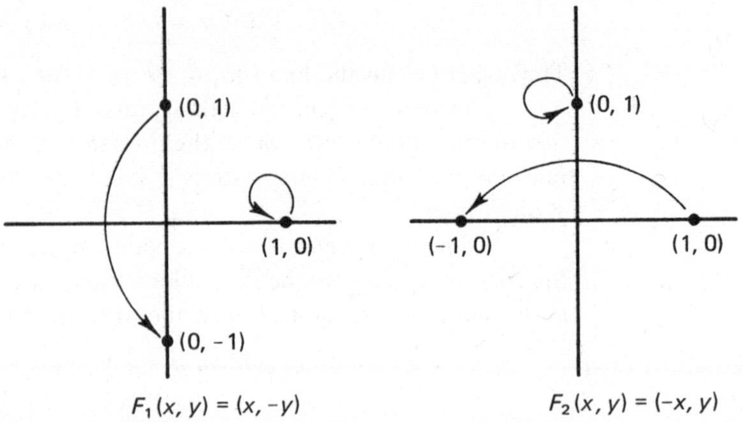

It is intuitively obvious that these transformations are all isometries. The question is "What properties of the matrices distinguish isometries from similarities and form other less special transformations?"

Similarities and Isometries

Let us now make a more systematic search for the conditions on the matrix

$$A = \begin{bmatrix} a_1 & b_1 \\ a_2 & b_2 \end{bmatrix}$$

that are necessary and sufficient to make the transformation a similarity. We note first a condition necessary for the right angle between the axes to be preserved.

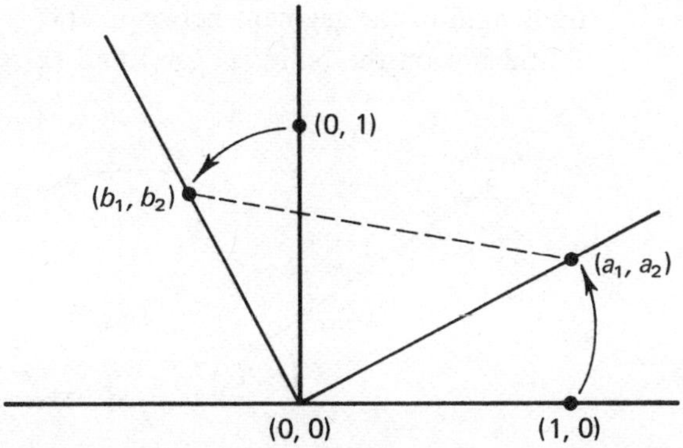

Since $T(1, 0) = (a_1, a_2)$ and $T(0, 1) = (b_1, b_2)$ and $T(0, 0) = (0, 0)$, it can be inferred from the Pythagorean law that the entries of A must satisfy

$$(a_1^2 + a_2^2) + (b_1^2 + b_2^2) = (b_1 - a_1)^2 + (b_2 - a_2)^2.$$

When this is simplified, the requirement becomes

$$a_1b_1 + a_2b_2 = 0. \tag{1}$$

Another necessary condition is evident from the diagram above. If A is the matrix of a similarity transformation, then the image distances must be some fixed multiple of the originals. The distances from (0, 0) to (1, 0) and to (0, 1) are both 1. Thus it is necessary that

$$\sqrt{a_1^2 + a_2^2} = \sqrt{b_1^2 + b_2^2} \quad \text{and} \quad a_1^2 + a_2^2 = b_1^2 + b_2^2. \tag{2}$$

If this is to be an isometry, then

$$\sqrt{a_1^2 + a_2^2} = \sqrt{b_1^2 + b_2^2} = 1 \tag{3}$$

must also be satisfied.

Surprisingly, these simple conditions on A are also sufficient to ensure that the transformation be a similarity and an isometry. In order that a transformation be a similarity, the length of every line segment must be magnified by the same amount, regardless of which points the segment joins.

Let (x_1, y_1) and (x_2, y_2) be any two points and denote the length of the line segment joining them by L. Then $L^2 = (x_1 - x_2)^2 + (y_1 - y_2)^2$. We want to show that conditions (1) and (2) on the entries of A will make the length of the segment between $T(x_1, y_1)$ and $T(x_2, y_2)$ depend only on L and not on the points (x_1, y_1) and (x_2, y_2).

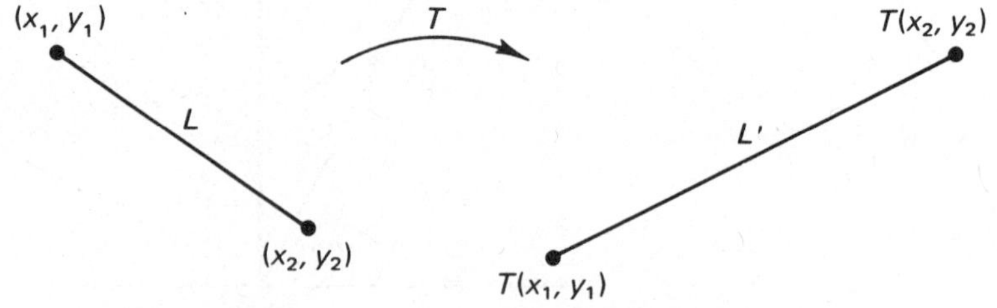

A short computation shows that $(L')^2$, the square of the distance between $T(x_1, y_1)$ and $T(x_2, y_2)$, is

$$(L') = (a_1^2 + a_2^2)(x_1 - x_2)^2 + (b_1^2 + b_2^2)(y_1 - y_2)^2 \\ + 2(a_1b_1 + a_2b_2)(x_1 - x_2)(y_1 - y_2).$$

If A satisfies conditions (1) and (2), then this equation becomes

$$(L')^2 = (a_1^2 + a_2^2)(x_1 - x_2)^2 + (b_1^2 + b_2^2)(y_1 - y_2)^2$$
$$(L')^2 = (a_1^2 + a_2^2)L^2 = (b_1^2 + b_2^2)L^2.$$

This last equation shows that L', the length of the image segment, is $\sqrt{a_1^2 + a_2^2}$ times the length of the original segment, no matter what the end points of the segment are. Such a mapping is a similarity transformation and will be an isometry if $a_1^2 + a_2^2 = 1$.

In short, we have found that a transformation T with matrix

$$A = \begin{bmatrix} a_1 & b_1 \\ a_2 & b_2 \end{bmatrix}$$

is a similarity transformation if and only if two conditions are satisfied.

1. $a_1b_1 + a_2b_2 = 0$
2. $a_1^2 + a_2^2 = b_1^2 + b_2^2$

The transformation is an isometry if, in addition, $a_1^2 + a_2^2 = 1$.

These conditions can be simplified. Consider what b_1 and b_2 must look like if a_1 and a_2 are known. The numbers b_1 and b_2 must then be solutions of the system of equations

$$a_1x + a_2y = 0$$
$$x^2 + y^2 = a_1^2 + a_2^2.$$

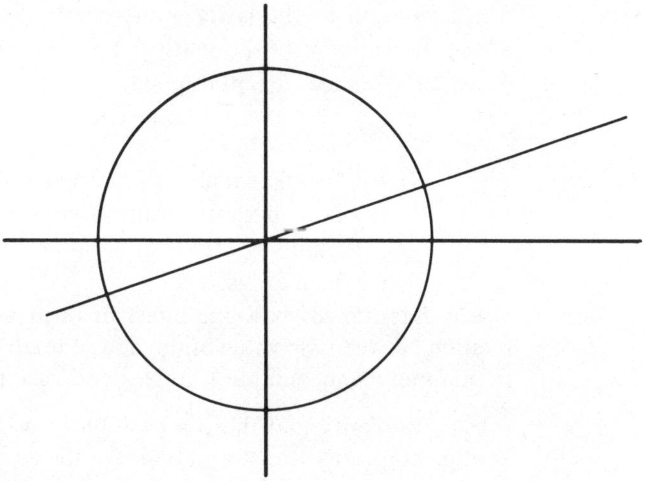

The graph of the first equation is a straight line that passes through the origin. The graph of the second equation is a circle with center at the origin and radius $\sqrt{a_1^2 + a_2^2}$. The graphs should intersect in exactly two points and the points should be diametrically opposite. A little more thought provides the two solutions $b_1 = a_2$ and $b_2 = -a_1$ or $b_1 = -a_2$ and $b_2 = a_1$.

Thus the matrix of a similarity is of one of the two forms:

$$\begin{bmatrix} a_1 & a_2 \\ a_2 & -a_1 \end{bmatrix} \quad \begin{bmatrix} a_1 & -a_2 \\ a_2 & a_1 \end{bmatrix}$$

and is an isometry if the determinant is -1 or $+1$.

Examples:

1. The mapping defined by $T_1(x, y) = (2x - y, x + 2y)$ is a similarity transformation in which every length is multipled by $\sqrt{5}$.

2. The mapping defined by

$$T_2(x, y) = \left(\frac{x}{2} + \frac{\sqrt{3}y}{2}, \frac{\sqrt{3}x}{2} - \frac{y}{2}\right)$$

is an isometry.

Summary

Linear transformations map straight lines into straight lines. It can be determined by inspection of the associated matrix whether or not a linear transformation is a similarity or an isometry. In particular, the absolute value of the determinant of the matrix of a similarity is the square of the factor by which distances are multiplied.

EXERCISES

1. Find some examples of matrices for transformations which are
   (a) Not similarity transformations.
   (b) Similarity transformations but not isometries.
   (c) Isometries.

Use a diagram to show the effect of the transformations. Check the correlation between the value of the determinant and the magnification of area by mapping some simple triangles and rectangles.

2. Verify that the composition of two similarity transformations is another similarity transformation. Do the same for the composition of two isometries. Show that the inverse of a similarity is a similarity and the inverse of an isometry is an isometry.

3. Make a table showing the results of combining the reflections F_1 and F_2 mentioned in this chapter. Extend the table as necessary to obtain a multiplication (composition) table for a set that is closed.

:wenty-four

abstract algebra

Abstract algebra is a branch of mathematics which has developed in the twentieth century as a natural consequence of the kinds of mathematical investigations which were conducted in the preceding one hundred years. Non-Euclidean geometries were invented about 1830 and it was realized that postulates other than those chosen by Euclid could lead to significant geometries. Gauss proved formally the facts about divisibility and studied the systems created by the definition of congruence. Lagrange, Abel, and Galois discovered the importance of certain sets of permutations while looking for methods of solving polynomial equations. Cayley introduced the notation for a matrix and started the investigation into rules for their manipulation. Dedekind and Cantor made their attempts to formally define the real number system and Peano tried to find axioms on which all mathematics could be based.

The attempts to provide a firm logical foundation for mathematics continued into the twentieth century. At the same time mathematicians became increasingly interested in the study of the structure of the systems that occur in the work of men such as Gauss, Galois, and Cayley. The properties of such seemingly different kinds of systems as the real numbers with addition, permutations with the composition operation, and matrices with multiplication had been seen to be strikingly similar.

Abstract algebra is the study of the properties of the different kinds of systems that have been found in algebra—without regard to the nature of the objects in the systems. In order to provide an indication of the way things are done in such investigations we shall discuss briefly one of the simplest of these systems.

Groups

We shall discuss the particular kind of system formally describe
by the following definition.

DEFINITION | A **group** G is a set on which is defined an operation that combines tw elements of the set such that

1. G is closed under the operation.
2. The operation is associative.
3. There exists an identity u for the operation.
4. Every element has an inverse with regard to the operation.

We have encountered several examples of groups. Some of them are liste below.

1. The set of complex numbers and the subsets of real number rational numbers, and integers, with the operation of addition.

2. The set of *nonzero* complex numbers and its subsets of rea numbers and rational numbers, with multiplication.

3. The set of classes of natural numbers, $\{\underline{0}, \underline{1}, \underline{2}, \ldots, \underline{m-1}\}$, wit addition defined modulo m.

4. The set S_n of all permutations of $\{1, 2, \ldots, n\}$ and some of th subsets of S_n, with the composition operation.

5. The set of four matrices with 0 and 1 entries, which are denote I, R_1, R_2, R_3 in Chapter 23, with the multiplication operation. Essentiall the same group is obtained by taking the set of associated transformation with the operation of composition of transformations.

6. The set of linear transformations of the plane and its subsets c similarity transformations and isometries, with the composition operatio Klein defined geometry as the study of *groups* of transformations (se p. 95). The sets of invertible matrices associated with these sets of tran formations form essentially the same groups with multiplication.

We shall next prove a few theorems about groups. The first thre contain routine facts that it is necessary to establish in order to be able t prove more significant results. They are included here mainly to provid an introduction to notations and methods of proof before the more difficu proof of an important theorem is discussed. The operation of a group wi be written like ordinary multiplication in order to simplify the notatio

THEOREM | A group contains exactly one identity element.

We know that there is at least one identity element u. It has the property that $gu = ug = g$ for every element g of the group. We must show that there is no other element with this property. Assume that an element x has the property; then $xu = u$. But since u is the identity, $xu = x$. Therefore, $x = u$.

THEOREM | Each element of a group has exactly one inverse.

Let g be an element of the group and let g^{-1} and v have the property of inverses. That is, $gg^{-1} = g^{-1}g = u$ and $gv = vg = u$. We must show that v and g^{-1} are the same element. From the fact that $gv = u$, it follows that $g^{-1}(gv) = g^{-1}u = g^{-1}$. But also $g^{-1}(gv) = (g^{-1}g)v = uv = v$. Thus, $g^{-1} = v$.

THEOREM | The inverse of gh is $h^{-1}g^{-1}$.

By the preceding theorem we need only show that $h^{-1}g^{-1}$ has the property that $(gh)(h^{-1}g^{-1}) = u$ and $(h^{-1}g^{-1})(gh) = u$. This is accomplished by using the associative law and the properties of u.

$$(gh)(h^{-1}g^{-1}) = g(hh^{-1})g^{-1} = (gu)g^{-1} = gg^{-1} = u$$
$$(h^{-1}g^{-1})(gh) = h^{-1}(g^{-1}g)h = h^{-1}(uh) = h^{-1}h = u$$

Lagrange's Theorem

We are now ready to consider a theorem which was first proved by Lagrange and which is sometimes called the First Fundamental Theorem about groups. As we have seen, Lagrange studied the sets of permutations that leave invariant certain combinations of the solutions of polynomial equations. These sets are probably the most important examples of groups. Cayley proved a theorem (the Second Fundamental Theorem) that states that *every* group is essentially the same as a group of "permutations" (one-to-one correspondences) of the elements of some set. The words *essentially the same* here mean that the only differences are in the terminology or notation; the elements and operation of the group can be renamed so that a "permutation" group is obtained. It is not surprising then that Lagrange discovered an important theorem about all groups by working with the subsets of the groups S_n that are themselves groups.

DEFINITION | A subset of a group G that is itself a group, using the same operation as G, is called a **subgroup of G.**

DEFINITION | If a group has n elements, where n is a natural number, then the group is called a **finite group of order *n*.**

LAGRANGE'S THEOREM | The order of a finite group is a multiple of the order of each of its subgroups.

The theorem says that if H is a subgroup of the group G, then the order of H divides the order of G. We shall prove this by showing that the elements of G can be divided into disjoint subsets in such a way that each subset contains exactly as many elements as does H. A few additional remarks about the examples in Chapter 20 will provide "concrete" models with which the proof of Lagrange's abstract theorem can be compared.

Example: The set $H = \{I, (123), (132)\}$ is the subgroup of S_3 that leaves $r_1^2r_2 + r_2^2r_3 + r_3^2r_1$ unchanged. The other three elements of S_3 form a set in which each element changes the expression in the same way. This set can be obtained by multiplying each element of H by any one of the elements not in H.

$$H_{(12)} = \{I(12), (123)(12), (132)(12)\} = \{(12), (13), (23)\} = H_{(13)} = H_{(23)}$$

The set S_3 is the union of the two disjoint subsets H and $H_{(12)}$.

Example: The set $H = \{I, (23)\}$ is a subgroup of S_3. If each element is composed with (12), which is not in H, then the set $\{(12), (132)\}$ is obtained. If H is multiplied by one of the remaining elements, then a third set of two elements is obtained. This uses up all the elements of S_3 and the order of S_3 is three times the order of H.

$$\begin{aligned} H = H_I = H_{(23)} &= \{I, (23)\} \\ H_{(12)} = H_{(123)} &= \{(12), (132)\} \\ H_{(13)} = H_{(132)} &= \{(13), (123)\} \end{aligned}$$

Example: The subgroup of S_4 that leaves $r_1 - r_2 - r_3 - r_4$ invariant can be used to partition S_4. The other subsets are obtained by multiplying each element of the subgroup by an element not in the subgroup.

$$\begin{aligned} H &= \{I, (34), (23), (234), (243), (24)\} \\ H_{(12)} &= \{(12), (12)(34), (132), (1342), (1432), (142)\} \\ H_{(13)} &= \{(13), (143), (123), (1423), (1234), (13)(24)\} \\ H_{(14)} &= \{(14), (134), (14)(23), (1234), (1324), (124)\} \end{aligned}$$

These sets have no elements in common and contain every element of S_4. The order of H divides the order of S_4 and the quotient is the number of sets.

These seemingly very special examples illustrate what Lagrange's theorem says happens in every finite group. The basic idea of the proof is that a subgroup can be used to partition the elements of the group into subsets having the same number of elements as the subgroup.

Proof of Lagrange's Theorem: Let G be a finite group and let H be a subgroup of G. For each element g of G, let Hg be the set of all elements obtained by multiplying each element of H (on the right) by the element g. The proof is conveniently broken up into two parts.

Lemma: Every set Hg contains exactly as many elements as does H.

We have only to show that no two different elements h_1 and h_2 of H give the same product when multiplied by g. Assume that $h_1g = h_2g$; then $(h_1g)g^{-1} = (h_2g)g^{-1}$ and $h_1(gg^{-1}) = h_2(gg^{-1})$ and $h_1u = h_2u$ and $h_1 = h_2$.

Lemma: If g_1 and g_2 are elements of G, then the sets Hg_1 and Hg_2 are either identical or disjoint.

We show that if the sets Hg_1 and Hg_2 have even one element in common, then they are identical. Let g belong to both sets. Then there are elements h_1 and h_2 in H such that $g = h_1g_1$ and $g = h_2g_2$. If hg_1 is any element of Hg_1, then

$$hg_1 = hug_1 = hh_1^{-1}h_1g_1 = hh_1^{-1}g = hh_1^{-1}h_2g_2 = (hh_1^{-1}h_2)g_2.$$

Since the product $hh_1^{-1}h_2$ of elements of H must be in H, we see that hg_1 must be in Hg_2. Thus $Hg_1 \subset Hg_2$. A similar argument shows that $Hg_2 \subset Hg_1$. Therefore, if any element g belongs to both sets, then $Hg_1 = Hg_2$.

The lemmas prove that the elements of G can be partitioned into the subsets of the form Hg. Each element belongs to exactly one subset and each subset contains exactly as many elements as does H. The number of elements in G is therefore the number of elements in H times the number of different subsets. *Q.E.D.*

Summary

Lagrange's Theorem is typical of the theorems of abstract algebra and, in fact, similar to many theorems of modern mathematics. It provides information about the structure of a variety of systems which contain many different kinds of objects but which have certain basic properties in common.

The development of abstract algebra was given impetus by the work of Galois. Galois showed the importance of the concept of a group by using it to extend the work of Abel on the solutions of polynomial equations. He investigated the groups of permutations of the solutions of polynomial equations. Galois discovered that a polynomial equation is algebraically solvable if and only if the subgroups of the group for the polynomial satisfy special conditions.

EXERCISES

1. Identify the identity element and the inverse of each element in the examples of groups listed on p. 216.

2. Prove that if g is an element of a group, then the inverse of g^{-1} is g. (See the theorems on inverses in this chapter.) Check the interpretation of this theorem in each of the examples of this chapter.

3. Prove that in a group there is exactly one solution of the equation $gx = h$, where g and h are elements of the group. (Show that it can be "solved" like an ordinary linear equation.) Verify this theorem in each of the examples.

DISCUSSION QUESTION

Find as many subgroups as possible in S_3 and S_4. Do you think that if a number d divides the order of a group, then there is a subgroup of order d?

part five

calculus

twenty-five

calculation of area and volume

The Greek mathematicians knew the formulas for finding the areas and volumes of the simple geometric objects that are bounded by straight lines and planes. They also were able to use their theory of ratios and commensurables to find formulas which show the relationship between areas of different conic sections and which relate the areas of conics to the areas of figures bounded by straight lines. This method could also be applied to find volumes.

The Greeks obtained their results by using a procedure called the *method of exhaustion.* European mathematicians began to make extensive use of a similar procedure based on the idea of *infinitesimals* more than three hundred years ago. Both of these methods are considered to be early forms of the modern calculus technique called *integration.*

The Method of Exhaustion

The method of exhaustion was probably first used by Eudoxus of Cnidas about 400 B.C. Euclid's *Elements* contains examples of its application; it is used to prove the second proposition in Book XII, which implies the existence of the number π.

Proposition (XII-2): The areas of circles are proportional to the square of the diameters.

The Greeks had already proved the analogous result for regular polygons; that is, they knew that the area of a regular n-sided figure is proportional to the square of any diagonal.

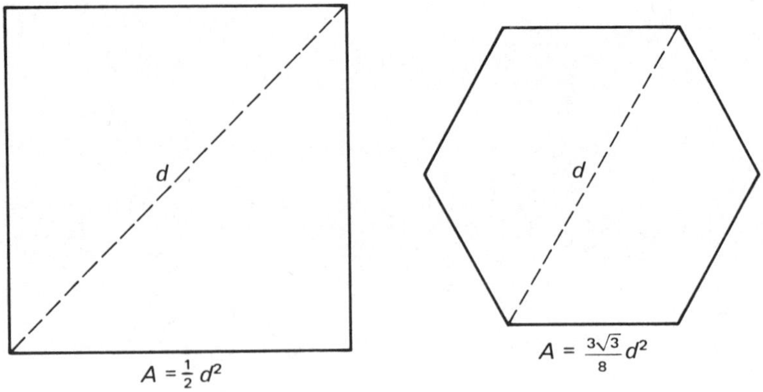

The present proposition says that there is some constant k (the number we now call $\pi/4$) such that for any circle $A = kd^2$. Euclid proved this by showing that for any two circles C_1 and C_2 with areas and diameters A_1, A_2, d_1, and d_2 the following equality holds.

$$\frac{A_1}{d_1^2} = \frac{A_2}{d_2^2}$$

The method is to show that the assumption that one of these quantities is larger than the other leads to a contradiction of the theorem for regular polygons.

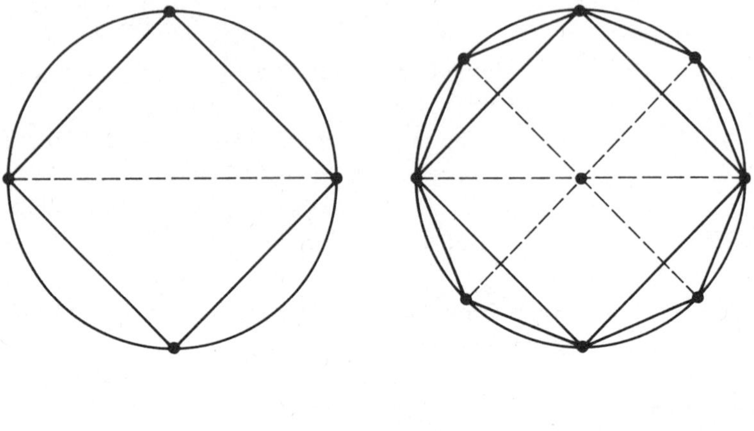

Assume that $A_1/d_1^2 > A_2/d_2^2$; then, $B/d_1^2 = A_2/d_2^2$ with $B < A_1$. Inscribe in C_1 a square with a diagonal on a diameter. The area of the square is more than half A_1 because it is exactly half of the square circumscribing the circle. Inscribe an octagon in C_1 by bisecting each side of the square. The additional area in the octagon, which is not in the square, is more than half of the part of the circle not in the square. (See the figure on the left below.)

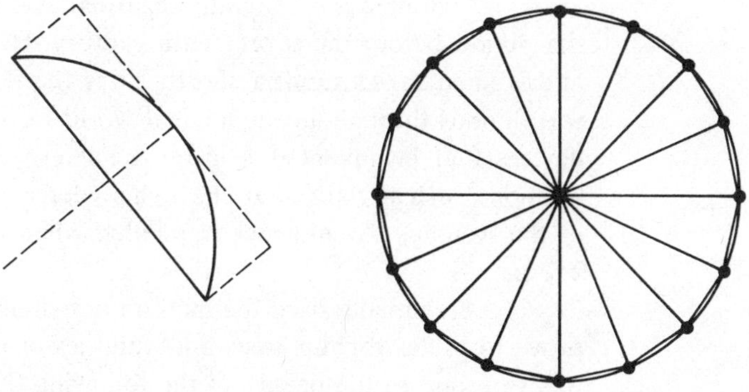

Thus the errors made in approximating the area of the circle by the areas of the square and octagon are, respectively, less than $A_1/2$ and $A_1/4$. The procedure of bisection to double the number of sides can be continued, and each time the amount of circle outside the new polygon will be less than half of what it was. Euclid now quotes his version of the Archimedean property of real numbers to claim that the approximation error can, in a finite number of steps, be made as small as desired. In particular, it can be made smaller than $A_1 - B$. When this is done, the area, call it P_1, of the polygon satisfies $B < P_1 < A_1$.

Inscribe in C_2 a similar polygon with area P_2; then

$$\frac{P_1}{d_1^2} = \frac{P_2}{d_2^2} \quad \text{or} \quad \frac{P_1}{P_2} = \frac{d_1^2}{d_2^2}$$

and therefore, using our hypothesis, we have

$$\frac{P_1}{P_2} = \frac{B}{A_2}.$$

Since $P_2 < A_2$, it must be that $P_1 < B$. This contradicts the fact that $B < P_1 < A_1$. The original assumption that $B < A_1$ must be false and the theorem is proved.

It is possible to obtain very good estimates for the value of the constant π by calculating the areas and perimeters of the polygons. For example, if a circle of radius 1 is chosen, then the area of any inscribed polygon is less than π and the area of any circumscribed polygon is greater than π. (The area of the circle is $\pi r^2 = \pi$.) Archimedes of Syracuse (287?–212 B.C.) used this kind of method to obtain the estimate $3\frac{10}{71} < \pi < 3\frac{1}{7}$. By the nineteenth century, methods based on this idea had been used to approximate π to several hundred decimal places.

Archimedes is generally regarded as the greatest mathematician of the period before the seventeenth century. He is thought to have visited and studied in Alexandria shortly after the death of Euclid. Archimedes was famous throughout the ancient world for the mechanical labor-saving devices that he invented. During the later years of his life he designed weapons, such as giant catapults, to help defend Syracuse against the attacks of the Romans. Archimedes was killed when the city fell to Marcellus in 212 B.C.

Archimedes used the method of exhaustion and great ingenuity to prove formulas for the areas and volumes of many geometric objects. He was so proud of the beauty of the following theorem that he had a representation of it carved on his tombstone.

THEOREM | A cylinder with base on a greatest circle of a sphere and with height equal to the diameter of the sphere has volume and surface area that are in the ratio of 3 to 2 with those of the sphere.

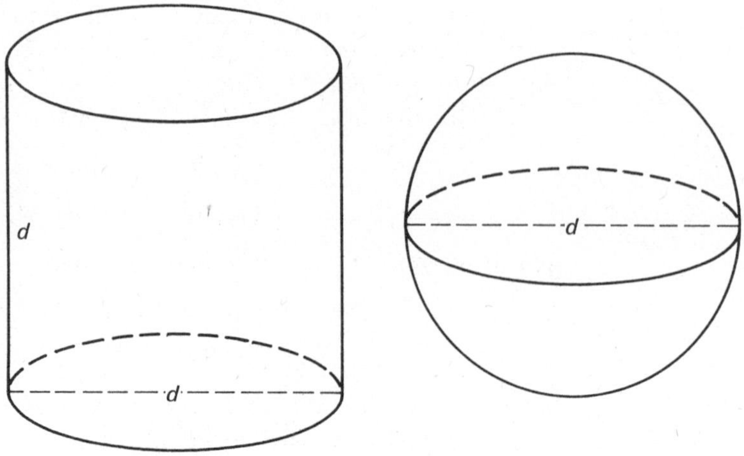

Archimedes proved this theorem with an elegant three-dimensional version of the proof used by Euclid for circles. His most famous and useful result with plane figures is the determination of the areas of segments of parabolas.

Proposition: The area between a parabola and one of its chords is equal to four-thirds the area of the inscribed triangle formed by using as the third vertex the point where the line through the midpoint of the chord, parallel to the axis of the parabola, intersects the parabola.

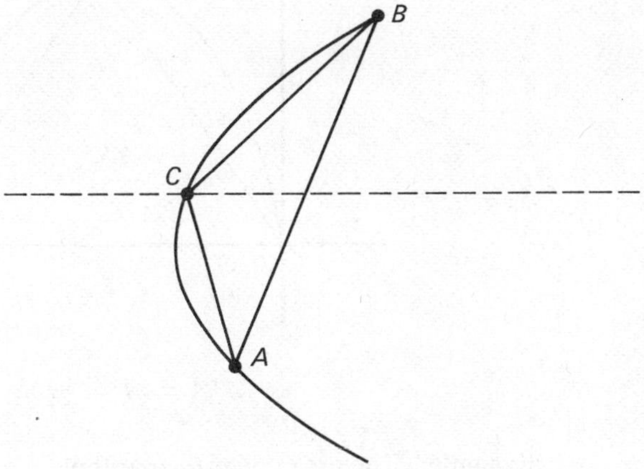

Archimedes showed that if the sides AC and BC are used as chords to form two new triangles in the same way, then the two triangles have a total area equal to one-fourth the area of triangle ABC. This procedure was continued and it was noted that each step used up more than half the remaining area. If T is the area of triangle ABC and S is the area of the segment of the parabola, then the estimates obtained in this way are

$$T < S < 2T$$
$$T(1 + \tfrac{1}{4}) < S < T(1 + \tfrac{2}{4})$$
$$T(1 + \tfrac{1}{4} + \tfrac{1}{16}) < S < T(1 + \tfrac{1}{4} + \tfrac{2}{16})$$
$$\vdots$$
$$T\left(1 + \frac{1}{4} + \frac{1}{16} + \cdots + \frac{1}{4^n}\right) < S < T\left(1 + \frac{1}{4} + \frac{1}{16} + \cdots + \frac{2}{4^n}\right)$$

By repeating the process enough times, each of the bounds can be made as close as one pleases to $\frac{4}{3}T$. Thus the assumption that S is larger or smaller than $\frac{4}{3}T$ can be contradicted and the theorem proved.

Using the analytic geometry that Descartes and Fermat develope for just this situation, we can apply the theorem to obtain numerical result

Example: The area under the parabola $y = 8x - x^2$ between (0, 0) an (8, 0) is $\frac{256}{3}$.

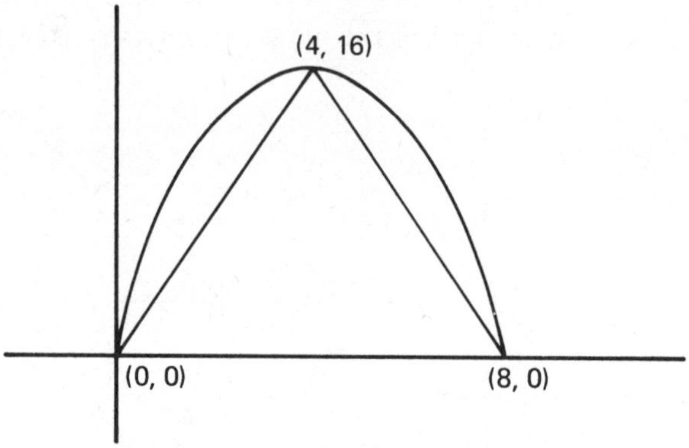

$$A = \tfrac{4}{3}T = \tfrac{4}{3}(\tfrac{1}{2}bh) = \tfrac{256}{3}$$

Example: The area under the parabola $y = x^2$ between (0, 0) and any poir $(x_1, 0)$ is the area of the large triangle minus four-thirds the area of th inscribed triangle. The area of each of these triangles can be calculated an the result is $A = \frac{1}{3}x_1^3$.

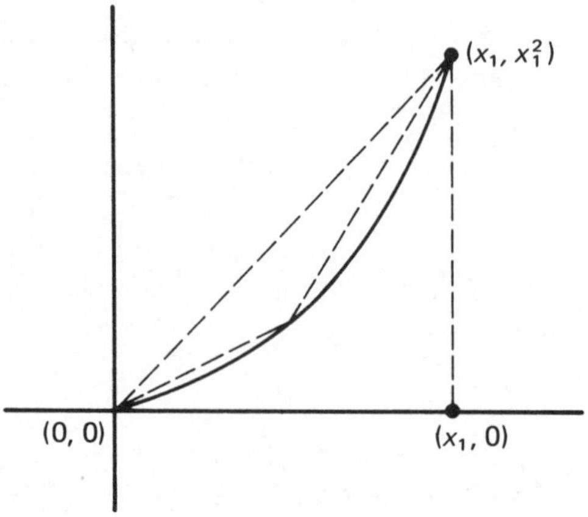

$$A = \tfrac{1}{3}x_1^3$$

In order to apply the method of exhaustion it is necessary to know the special properties of the curve involved and to use the Archimedean property of the real numbers in some form. The basic idea of "exhaustion" is simple, however. The properties of the curve are used to obtain a sequence of lower bounds and a sequence of upper bounds for the size of a quantity (such as area or volume). The bounds become arbitrarily close to some number. Any assumption that the size of the quantity is different from the number leads to the contradictory conclusion that it is outside the bounds.

Infinitesimals

By the time translations of Archimedes' work were printed in Europe in the sixteenth century, a very similar method was already in use. The results were accepted as valid, however, without going through the "details" of proof by contradiction. Various philosophical theories on the nature of matter in the real world were quoted to explain why the method should work. Most of these theories contain the idea, in one form or another, that matter is composed of infinite numbers of infinitesimal pieces.

In the early part of the seventeenth century Galileo Galilei (1564–1642) and Johann Kepler (1571–1630) made their investigations of the laws of motion. Kepler's principle interests involved the motion of planets. He had inherited the meticulous records of astronomical observations made over a period of twenty years by his predecessor Tycho Brahe. Kepler studied this data to determine the orbits of the heavenly bodies. He was also concerned about more down-to-earth subjects and contributed some new geometric formulas in his *Nova Stereometria doliorum vinariorum* (New Volume Measurements for Wine Barrels).

Kepler's reasoning was based on assumptions such as the one that ellipses can be partitioned into an infinite number of triangles with infinitesimal area. The diagram below is an attempt to illustrate this idea.

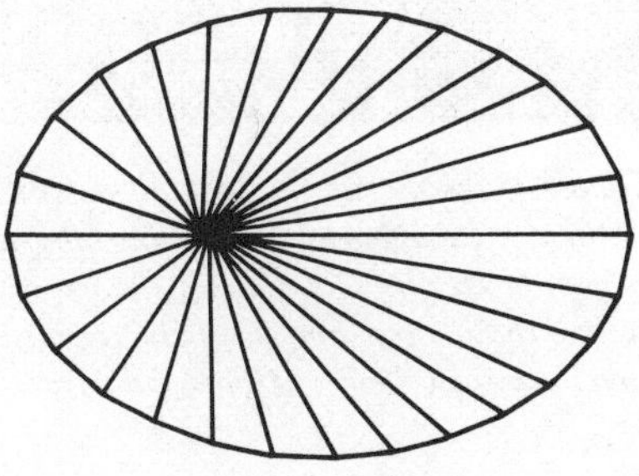

Galileo Galilei. *Galileo achieved fame because of his mathematical theories explaining physical phenomena. His attempts to popularize the Copernican theory of the universe led to his trial by the Inquisition and he spent the last years of his life under house arrest.* (*George Arents Research Library*)

It must be imagined that the radii from the focus are so close together that the triangles have areas too small to be measured.

Galileo and his students Bonaventura Cavalieri (1598–1647) and Evangelista Torricelli (1608–1647) found significant new applications of methods for finding areas. Galileo discovered the rules that govern the motion of freely falling bodies. Experimental evidence showed him that the acceleration of such objects is constant and that velocity is proportional to elapsed time. Furthermore, if the constant of acceleration is g, then $v = gt$. He also discovered that the distance traveled in any amount of time t_1 is $\frac{1}{2}gt_1^2$. Diagrams such as the one below made this seem reasonable.

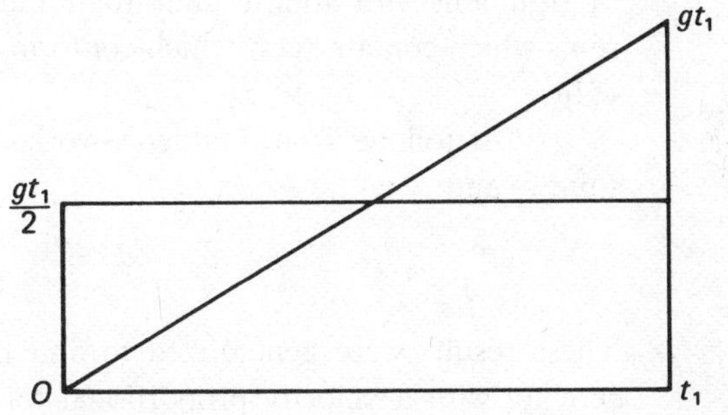

Galileo interpreted the area of the large triangle as the distance traveled and noted it was the same as the area of the rectangle, which represents the distance traveled in time t_1 at the constant average velocity of $\frac{1}{2}gt_1$.

When translated into modern notation and terminology, Galileo's proof is as follows: The velocity is a function of time and the graph of the function $v = gt$ is a straight line through the origin.

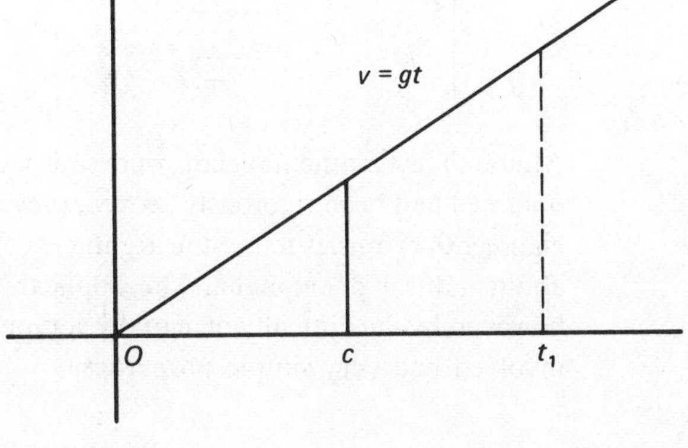

It is clear that for a body moving at a *constant* velocity the distance traveled is the product of velocity with time. If c is any instant of time, then the velocity at that time is "constant." Therefore, the area *covered by* the line segment at c, which has length equal to the "constant" velocity, is equal to the infinitesimal amount of distance traveled in the infinitesimal amount of time represented by the *width* of the line. Since the triangle under the curve $v = gt$ is the "sum" of all such line segments, the area of the triangle is the total distance covered.

Galileo's result is true, but his argument was criticized by philosophers and theologians as well as mathematicians. The usefulness of the result outweighed these objections. In the following two hundred years this type of argument with infinitesimal quantities was used to "prove" many theorems which contain facts which conform to reality and have great practical value.

It follows from Galileo's work that the distance covered between time t_1 and t_2 is

$$d_2 - d_1 = \tfrac{1}{2}gt_2^2 - \tfrac{1}{2}gt_1^2$$

These results were generalized to find the distance covered by an object moving with a velocity proportional to higher powers of time. The table below shows the results of the investigations of Torricelli.

v	*Distance:* $t = 0$ *to* $t = t_1$	*Distance:* $t = t_1$ *to* $t = t_2$
k	kt_1	$kt_2 - kt_1$
kt	$\frac{1}{2}kt_1^2$	$\frac{1}{2}kt_2^2 - \frac{1}{2}kt_1^2$
kt^2	$\frac{1}{3}kt_1^3$	$\frac{1}{3}kt_2^3 - \frac{1}{3}kt_1^3$
$\vdots$	$\vdots$	$\vdots$
kt^n	$\dfrac{1}{n+1}kt_1^{n+1}$	$\dfrac{1}{n+1}kt_2^{n+1} - \dfrac{1}{n+1}kt_1^{n+1}$

Although few mathematicians were satisfied that these and other results they obtained had been rigorously proved, they continued in the seventeenth and eighteenth centuries to use the arguments to find and investigate useful facts about natural phenomena. The applicability of the method was limited, however, by the difficulty of actually computing the bounds unless the curves involved had very simple properties.

Summary

The Greeks used the method of exhaustion to obtain formulas for areas and volumes and proved that the results are correct. During the Renaissance, European mathematicians developed a similar method and applied it to discover and explain the natural laws about moving bodies. Even though there were many who doubted that the method was logically sound, it became widely used because the results obtained were consistent with their observations of natural phenomena.

EXERCISE

Calculate the areas under the following curves between the indicated values of x by using Torricelli's formula and some common sense. Check your results graphically.

1. $y = x$ between $x = 1$, $x = 3$.
2. $y = 2x$ between 3 and 6.
3. $y = x^4$ between 0 and 3 and between -2 and 2.
4. $y = x + 4$ between 1 and 5.
5. $y = 3x^2 + 1$ between 0 and 2.

DISCUSSION QUESTION

Try to develop a systematic method by which areas and volumes bounded by curves could be calculated in practical problems and which takes advantage of the analytic geometry representation of the curves. That is, choose a simple function of x such as $f(x) = x^2$ and explain a systematic procedure for applying the method of exhaustion to find the area under the curve $y = f(x)$.

twenty-six

tangents, maxima, and minima

Pierre de Fermat developed analytic geometry so that he coul investigate the properties of curves such as the conic sections. He wa particularly interested in being able to find the lines tangent to these curves The slope of the tangent indicates the "slope" of the curve and muc information about the behavior of a function can be obtained from knowledge of the tangents to the graph. In particular, the points where th curve achieves a maximum or minimum value as a function of x must hav tangents with slope zero.

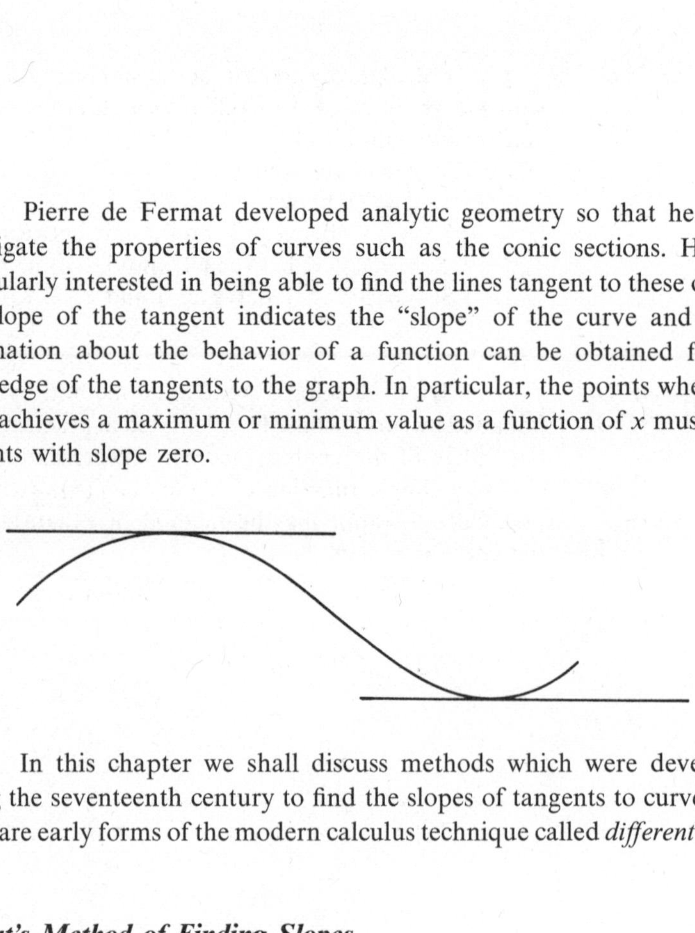

In this chapter we shall discuss methods which were develope during the seventeenth century to find the slopes of tangents to curves an which are early forms of the modern calculus technique called *differentiation*

Fermat's Method of Finding Slopes

Since most of Fermat's mathematical work was published posthu mously, it is difficult to determine the order in which he made his discoveries

Apparently he first wrote *Method for Finding Maxima and Minima* and then generalized the method so that it could be used to find all tangents. The problem that follows is like one that was used by Fermat to demonstrate the power of his procedure.

Problem: To divide a segment of length b into two pieces so that the product of the pieces is a maximum.

If one of the pieces is chosen to be of length x, then the other is $b - x$ and the product is $x(b - x)$. The problem then is to find the value of x which makes $f(x) = x(b - x)$ maximal. If this function is graphed, the curve is a parabola.

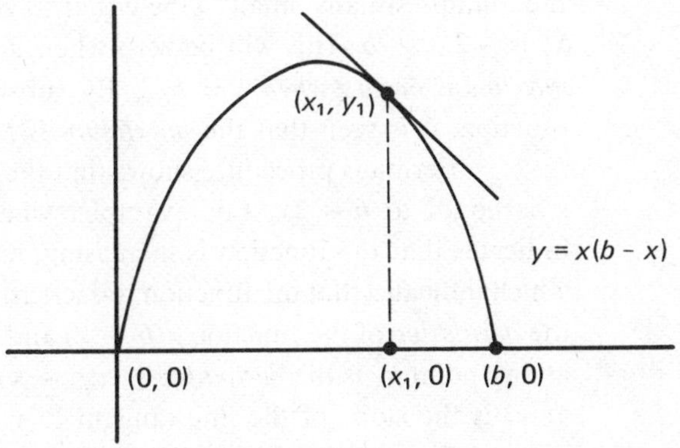

The slope of the tangent to the curve at any point (x_1, y_1) can be approximated by the slope of the chord drawn to a nearby point. If the other point has first coordinate $x_1 + e$, then the smaller that e is chosen, the better the estimate should be.

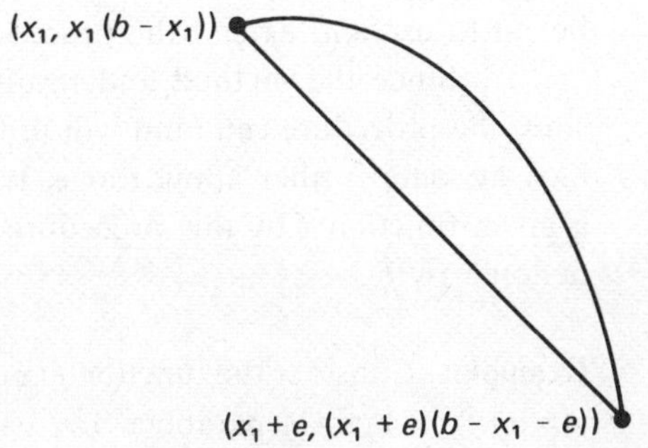

The slope of the chord between these two points on the curve is

$$\frac{(x_1 + e)(b - x_1 - e) - x_1(b - x_1)}{(x_1 + e) - x_1}.$$

When this is simplified, the result is

$$\frac{-2x_1e + be + e^2}{e}.$$

Dividing by e yields

$$-2x_1 + b + e.$$

Fermat then claims that since the best estimates are given whe e is very small, the slope of the tangent at x_1 can be obtained by lettin the "infinitesimally small" e be equal to zero. Thus the slope at any poin x_1 is $-2x_1 + b$. This will be zero when $y = f(x)$ is a maximum. Thus th *maximum occurs when* $x = b/2$. By substituting this value of x into th function, it is seen that the *maximum is* $f(b/2) = b^2/4$.

Fermat's procedure shows that the slope of the tangent at any poin x is equal to $b - 2x$. For example, when $x = 0$, the slope is b, whic indicates that the function is increasing, and when $x = b$, the slope is $-b$ which indicates that the function is decreasing. The function $b - 2x$ is calle the *derivative* of the function $x(b - x)$ and the value of the function $b - 2x$ at any point x_1 is the *derivative of* $x(b - x)$ *at* x_1. In this case the derivativ at x_1 is the slope of the line tangent to $y = f(x)$ when $x = x_1$.

Fermat's "proof"—like Galileo's—has a great deal of intuitiv appeal, especially when a graph is drawn to aid the imagination. Objection were raised concerning the use of the "infinitesimal number" e. It is assume nonzero long enough to divide by it and then called zero to obtain th answer. Such arguments were known to sometimes lead to false results. Th answers obtained by Fermat, however, when he applied the method to fin tangents to simple curves, always seemed to be correct and mathematician began to use and extend the procedure.

Since the method and results are important in mathematics an since the procedure can (and will in Chapter 28) be made precise, we shal look at some further applications. In general, the function obtained fron a given function f by this procedure will be called the *derivative of* f an denoted by f'.

Example: Consider the function $f(x) = x^2$. When this is graphed, the curv $y = f(x)$ is again a parabola. Let us find the slope of the tangent at an point by calculating the derivative. If x_1 is any value of x, then the slope of the chord between the points of the curve where $x = x_1$ and $x = x_1 + e$ is

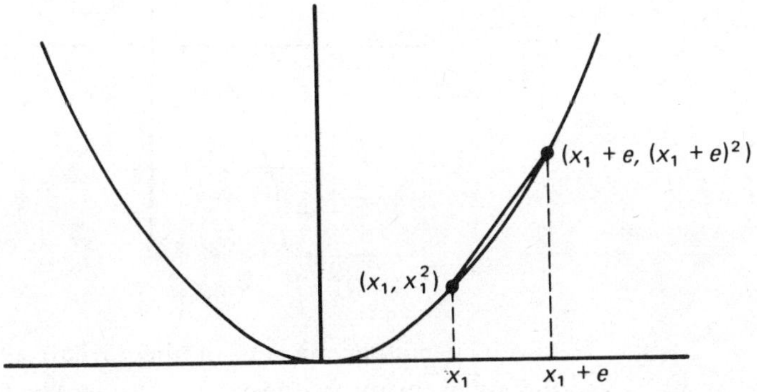

$$\frac{f(x_1 + e) - f(x_1)}{(x_1 + e) - x_1} = \frac{(x_1 + e)^2 - x_1^2}{e} = 2x_1 + e.$$

By "letting e become zero," or "in the limit," the slope at x_1 is $2x_1$. In general, the slope at any point x is $2x$, and we have $f'(x) = 2x$. Notice that the derivative (slope) is zero when $x = 0$ and the function has a minimum value there. Also the slope is negative when x is negative and positive when x is positive.

Example: Consider the function $f(x) = x^3$. The derivative at a point x_1 should be "the limit when e becomes 0" of the ratio

$$\frac{f(x_1 + e) - f(x_1)}{e} = \frac{(x_1 + e)^3 - x_1^3}{e} = 3x_1^2 + 3x_1e + e^2.$$

We conclude therefore that $f'(x) = 3x^2$. When $x = 1$, the value of y is also 1 and the slope at (1, 1) is $f'(1) = 3$. The slope of the tangent is also 3 at the point $(-1, -1)$. The slope at (2, 8) is 12 and it is also 12 at $(-2, -8)$. The slope is 27 at (3, 27) and at $(-3, -27)$. The slope at $(\frac{1}{2}, \frac{1}{8})$ and $(-\frac{1}{2}, -\frac{1}{8})$ is $\frac{3}{4}$. The slope of the curve $y = f(x)$ is zero only when $x = 0$ and is never negative. Thus y never decreases when x increases and there are no maximum or minimum points. When information of this kind is plotted on a coordinate system, it can be seen that the graph of the function $y = x^3$ is like that shown on the next page.

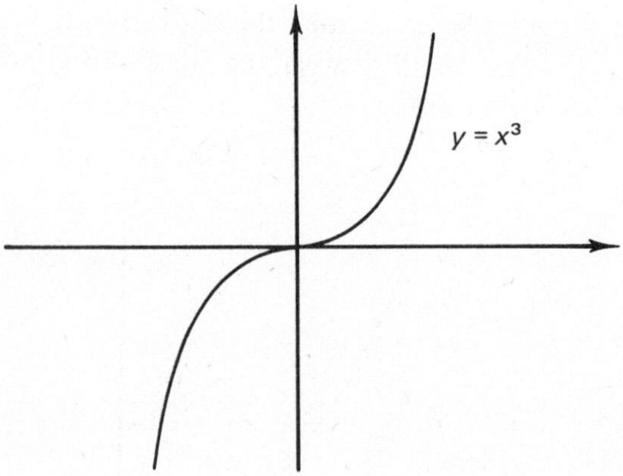

The mathematicians of the seventeenth century recognized the usefulness of this procedure and soon discovered how it can be applied to solve practical problems. They were severely hampered, however, by the difficulty of calculating the value of the derivative in any except the most elementary situations. Some indication of the difficulty can be obtained by considering what must be done to compute the slope of the tangent to a semicircle by this method.

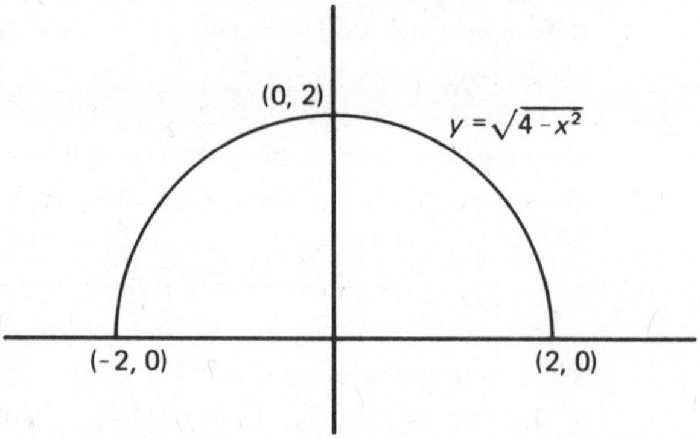

Here the ratio that must be considered is

$$\frac{f(x_1 + e) - f(x_1)}{e} = \frac{\sqrt{4 - (x_1 + e)^2} - \sqrt{4 - x_1^2}}{e}.$$

Barrow's Differential Triangle

Isaac Barrow (1630–1677) gave a method of finding the slopes of the tangents to more complicated curves. Barrow studied mathematics and divinity at Cambridge. After traveling in the Middle East he was ordained and he taught mathematics at Cambridge. In 1669 Barrow gave up mathematics and devoted full time to divinity. His student, Isaac Newton, was appointed to his place as professor of mathematics.

One of Barrow's last acts as a mathematician was to publish his *Lectiones opticae et geometricae*. In one of the lectures on geometry he gave the method of finding the slope of tangents of complicated curves by using what became known as *Barrow's differential triangle*. He gave the method without proof and showed several examples of applications. His method is very similar to a technique that was proved valid in the nineteenth century and is used today.

Barrow did not use analytic geometry, but he did use diagrams and his method can be readily translated into modern notation. (We shall, however, adhere to Barrow's explanations.) Given a curve and a value of x at which the slope of the tangent is desired, choose a value $x + e$ that is "indefinitely" close to x. Let a be the difference in the values of y at these points.

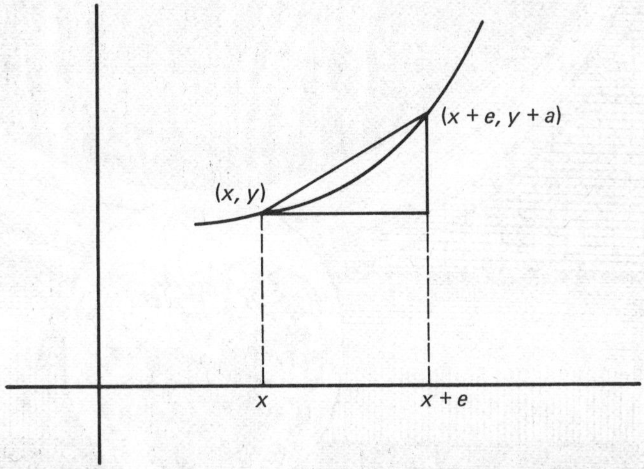

The method then is to write the equation showing that $(x + e, y + a)$ is on the curve and observe the following rules.

Isaac Barrow. *Professor of geometry at Cambridge, Barrow lectured on the methods which were later extended by Newton and are now called calculus. (George Arents Research Library)*

1. Omit all terms that contain a or e to a power higher than the first (because they don't amount to anything).

2. Omit all terms that do not contain a or e (because they will cancel out anyway).

3. Solve for a/e, which will be the value of the slope.

Example: A simple demonstration of the power of the method can be given by reconsidering the circle $x^2 + y^2 = 4$.

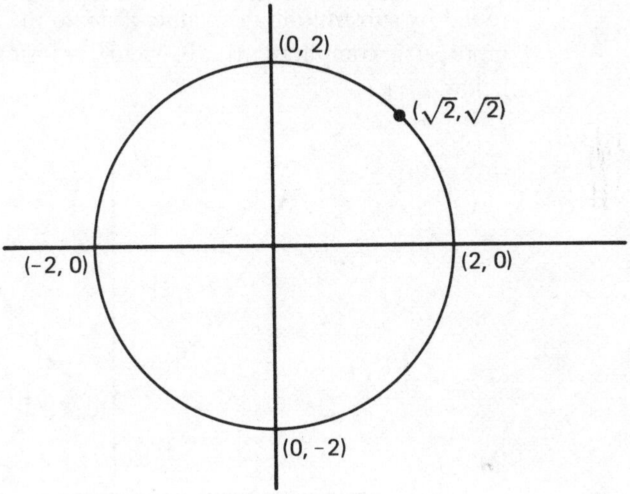

$$(x + e)^2 + (y + a)^2 = 4$$
$$x^2 + 2xe + e^2 + y^2 + 2ya + a^2 = 4$$
$$x^2 + 2xe + y^2 + 2ya = 4 \qquad \text{(Rule 1)}$$
$$2xe + 2ya = 0 \qquad \text{(Rule 2)}$$
$$\frac{a}{e} = \frac{-x}{y} \qquad \text{(Rule 3)}$$

This formula for the slope shows that the slope is zero at $(0, \pm 2)$, is ± 1 at $(\pm\sqrt{2}, \pm\sqrt{2})$, and is undefined at $(\pm 2, 0)$.

Example: We can find information about the graph of the hyperbola $y^2 - 4x^2 = 48$ by applying Barrow's method to calculate the slopes of the tangents to the curve. The derivative (of y with respect to x) is $4x/y$ since

$$
\begin{aligned}
(y+a)^2 - 4(x+e)^2 &= 48 \\
y^2 + 2ay + a^2 - 4x^2 - 8ex - 4e^2 &= 48 \\
y^2 + 2ay - 4x^2 - 8ex &= 48 \qquad \text{(Rule 1)} \\
2ay - 8ex &= \ \ 0 \qquad \text{(Rule 2)} \\
\frac{a}{e} &= \frac{4x}{y} \qquad \text{(Rule 3)}
\end{aligned}
$$

Thus the hyperbola has horizontal tangents if and only if $x = 0$. By subs tution we find that when $x = 0$ the values of y are $4\sqrt{3}$ and $-4\sqrt{3}$, ar $(0, 4\sqrt{3})$, $(0, -4\sqrt{3})$ are points on the curve. The tangents have slope equ to 1 if and only if $y = 4x$. The corresponding points on the curve can l found by substituting this value of y into the equation for the curve. A fe more such computations will yield the information shown on the grap below.

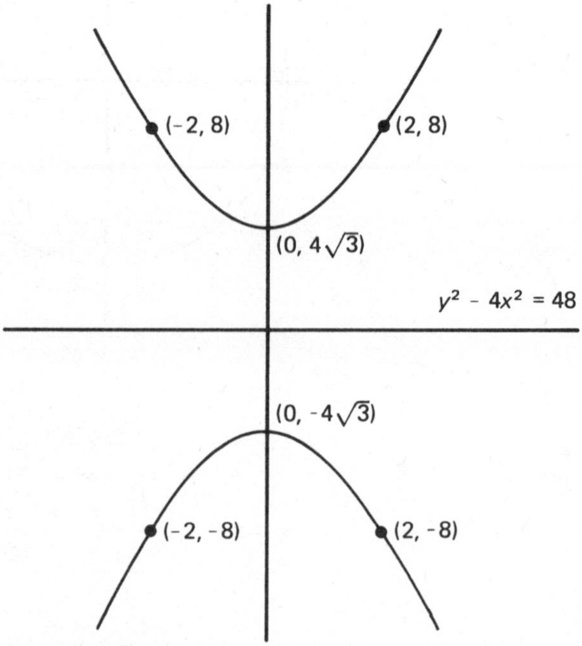

Barrow's method is similar to the procedure used by Fermat. I principle advantage is that the values of y $[= f(x)]$ need not be given e plicitly in terms of x. The modern method is in fact called *implic* differentiation.

Summary

Important properties of a curve can be discovered from a consideration of the slopes of the lines tangent to the curve. Fermat used analytic geometry to develop a method for finding the slopes of tangent lines. Barrow found a technique for finding the tangents to curves with more complicated equations.

Barrow saw a connection between these methods and those used by men such as Galileo and Kepler for finding areas and volumes. In the next chapter we shall discuss the work of his successors who "invented" calculus by explaining how these methods are related and utilizing this relationship.

EXERCISES

1. Use Fermat's method to find the derivative of each of the following functions. Check your work by considering the graphs of the curves $y = f(x)$.

$$f(x) = 6x \qquad f(x) = x^2 + 9$$
$$f(x) = 9 \qquad f(x) = x^2 + 6x$$
$$f(x) = 6x + 9 \qquad f(x) = x^2 + 6x + 9$$

2. Use Barrow's method to find the derivatives of the functions $x(b - x)$, x^2, and x^3. Compare your work and results with those obtained by Fermat's method.

3. One hundred feet of fence is to be used to enclose a rectangular plot. What should be the dimensions of the plot in order that it have maximum area?

DISCUSSION QUESTIONS

1. Use one of our "working definitions" of the derivative to determine the derivative of the function $f(x) = x^n$ for $n = 1, 2, 3, \ldots$. Guess the general rule, and try to explain how it could be proved.

2. Investigate the conjecture that "the derivative of the sum of two functions is the sum of the derivatives of the functions." (See the experimental evidence provided by Exercise 1.) Try to find an informal proof of the conjecture or an example in which it is false.

3. Reconsider the method of double false position mentioned in Chapter 28. Note also the second discussion question of that chapter.

twenty-seven

calculus

Kepler published his major works, the *Astronomia Nova* and *Harmonice Mundi,* in the years 1609 and 1619, respectively. On the basis of many years of careful observations he established the laws of planetary motion that bear his name:

1. The orbit of every planet is an ellipse with the sun at a focus.

2. A line from the sun to a planet sweeps out equal areas in equal times.

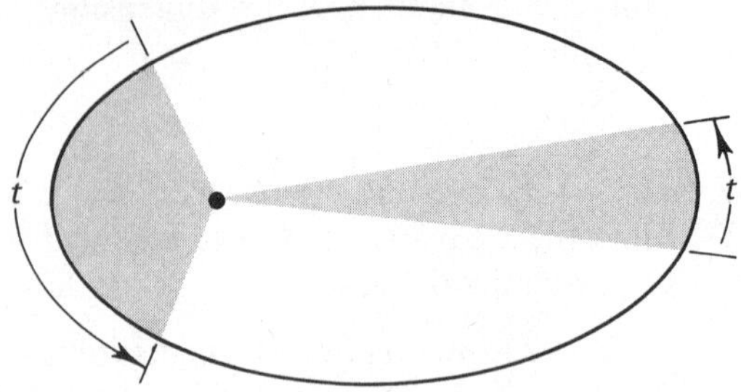

3. The square of the period of revolution is proportional to the cube of the mean distance of the planet from the sun.

These laws were empirical. Kepler found, after much arduous computation, that his experimental data supported these general statements. Newton's attempts to explain and prove these laws led him to the discovery of the basic rules of calculus.

Newton and Leibniz

From about 1665 to 1685 the great English mathematician Isaac Newton (1642–1727) studied Kepler's Laws and the mathematical techniques we have described in the last two chapters. Newton was a student under Barrow at Cambridge. After completing his undergraduate work, he returned to his farm for two years while the universities were closed because of the Great Plague. He returned to Cambridge to continue his experiments with optics and chemistry and worked on the mathematical theories that he had developed during his leave. In 1669 he succeeded Barrow as professor and taught for more than thirty years. Later he was called to supervise the issuance of new coinage, was appointed master of the mint, and became an active member of Parliament. Newton continued his research in the natural sciences and theology until his death.

By 1687, when he published the *Philosophiae Naturalis Principia Mathematica,* Newton had made several momentous discoveries. In addition to his basic laws of motion he postulated the universal theory of gravitation: Any two particles of matter in the universe attract each other with a force that is inversely proportional to the square of the distance between them. Moreover, Newton was able to *prove* Kepler's Laws from this postulate.

To accomplish these proofs, Newton applied the methods that he called the "direct and inverse method of fluxions." The word *fluxion* is Newton's term for the derivative and he called the infinitesimal quantities, such as those used by Fermat and Barrow, the *moments of fluxions*. He chose this terminology because he viewed curves as a "steady flow of points" rather than a static collection of infinitesimals. He interpreted the derivative, which had previously been thought of as the slope of the tangent line, as the instantaneous rate of change of y with respect to x. Newton's physical interpretation of the process is probably due in part to the fact that his most important applications were to moving objects.

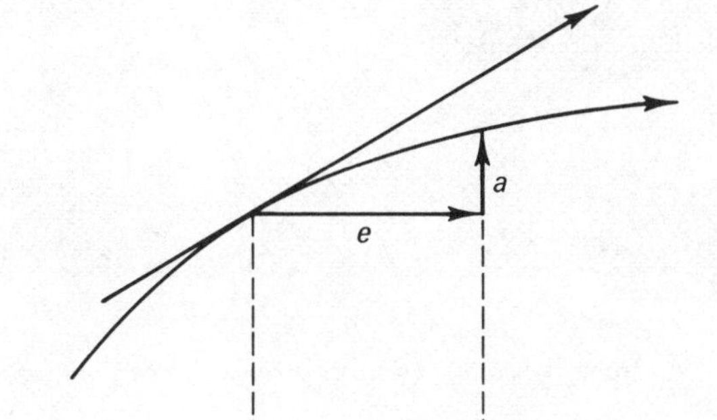

Isaac Newton. *(George Arents Research Library)*

$$\text{Average rate of change} = \frac{\text{change in } y}{\text{change in } x} = \frac{a}{e}$$

$$\text{Instantaneous rate of change} = \text{limit as } e \text{ goes to zero of } \frac{a}{e}$$

The "direct method of fluxions" is Newton's refinement of the procedure for finding derivatives. The "inverse method" is like the procedure used by Archimedes and Galileo for finding areas and distances. Before explaining why Newton thought these two methods to be inverses of each other, it is convenient to mention another major contributor to the development of the calculus.

Gottfried Wilhelm von Leibniz (1646–1716), the German philosopher, diplomat, and mathematician, was also working with these two methods during this period. Leibniz studied law and served as a diplomatic advisor when a young man. For most of his life he supported himself by serving as advisor and historian to the noble Hanover family. One of those he served became King George I of England.

Among Leibniz' goals was the establishment of a symbolic language in which all scientific topics, including logic and philosophy, could be formally discussed. When he learned of the difficulties that mathematicians were having with infinitesimals, he developed a notation by which the procedures could be conveniently expressed. The use of these symbols allowed Leibniz to manipulate the quantities in ways that facilitated computation and led him to the discovery of new general rules. The British, who became outraged because they believed that Leibniz was claiming and receiving credit for discoveries made by Newton, fell behind in mathematical research in the eighteenth century because they refused to make use of Leibniz' notation and other advances made by mathematicians on the Continent.

Leibniz and Newton probably both deserve the credit for originating the systematic study of the two techniques that are now called the *differential calculus* and the *integral calculus* and for developing many systematic procedures for their application. Leibniz introduced a notation for the procedure he used to solve the "inverse tangent" problem, which is now called *integration.* Since he thought of this process as the "limit of sums," he chose the old form of the letter *s*, ∫, as the symbol. That is, he represented the *result* of using a procedure like the method of exhaustion to find the area under the function $f(x)$ between the points a and b by

$$\int_a^b f \quad \text{or} \quad \int_a^b f(x)\,dx.$$

The symbol is called the *integral sign* and the notation is read "the integr of *f* from *a* to *b*." Although the second notation is more suggestive and mo widely used today, we shall elect the simpler one.

The Fundamental Theorem of Calculus

Using the notation of Leibniz we can compile some of the simp results of differentiation (finding the derivative) and integration that we mentioned in preceding chapters. An appropriate arrangement of th information will make it easier to understand why the mathematicians can to view differentiation and integration as inverse methods.

f'	f	$\int_0^u f$	$\int_a^b f$
1	x	$\frac{1}{2}u^2$	$\frac{1}{2}b^2 - \frac{1}{2}a^2$
$2x$	x^2	$\frac{1}{3}u^3$	$\frac{1}{3}b^3 - \frac{1}{3}a^2$
$3x^2$	x^3	$\frac{1}{4}u^4$	$\frac{1}{4}b^4 - \frac{1}{4}a^4$
nx^{n-1}	x^n	$\frac{1}{n+1}u^{n+1}$	$\frac{1}{n+1}b^{n+1} - \frac{1}{n+1}a^{n+1}$

Consider the first three columns. The first and third columns conta the results of differentiating and integrating the functions in the secor column. On the other hand, the functions in the second column can obtained by integrating the functions in the first column. Furthermore, t functions in the second column are the derivatives of those in the thi column. When these observations are combined with the information co tained in the fourth column and generalized, we have the most importa theorem about these two techniques.

FUNDAMENTAL THEOREM OF CALCULUS

If f is a function defined on the interval $[a, b]$ such that

$$\int_a^b f$$

exists and if F is a function defined on $[a, b]$ such that $F' = f$, then

$$\int_a^b f = F(b) - F(a).$$

The abstract nature of the statement of this theorem is a measu of its great generality. It gives a formula for the integral over an interv

Gottfried Wilhelm von Leibniz.
(*George Arents Research Library*)

of *any* function that is integrable and is the derivative of some other function The theorem applies to functions that are extremely complicated and provides a method for finding the result of integrating them over any interval. All that has to be done is to compute the difference between the values of this other function at the ends of the interval.

It would be impossible in our brief look at the calculus to show the extensive consequences of this theorem and we shall have to conten ourselves with a pair of simple applications.

Example: Since the derivative of the function $F(x) = x^3$ is $f(x) = 3x^2$, we can calculate the area under the curve $y = f(x)$ between any two point very easily.

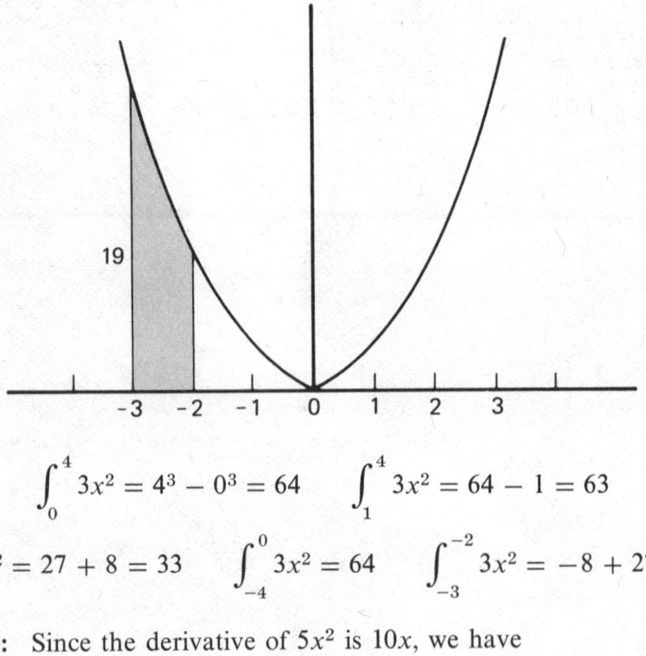

$$\int_0^4 3x^2 = 4^3 - 0^3 = 64 \qquad \int_1^4 3x^2 = 64 - 1 = 63$$

$$\int_{-2}^3 3x^2 = 27 + 8 = 33 \qquad \int_{-4}^0 3x^2 = 64 \qquad \int_{-3}^{-2} 3x^2 = -8 + 27 = 19$$

Example: Since the derivative of $5x^2$ is $10x$, we have

$$\int_1^6 10x = 5(6)^2 - 5(1)^2 = 175 \qquad \int_2^6 10x = 5(6)^2 - 5(2)^2 = 160$$

$$\int_1^7 10x = 5(7)^2 - 5(1)^2 = 240 \qquad \int_0^8 10x = 5(8)^2 - 5(0)^2 = 320.$$

The definitions of the integral and the derivative given by Newton and Leibniz were not significantly more precise than those used earlier. Th

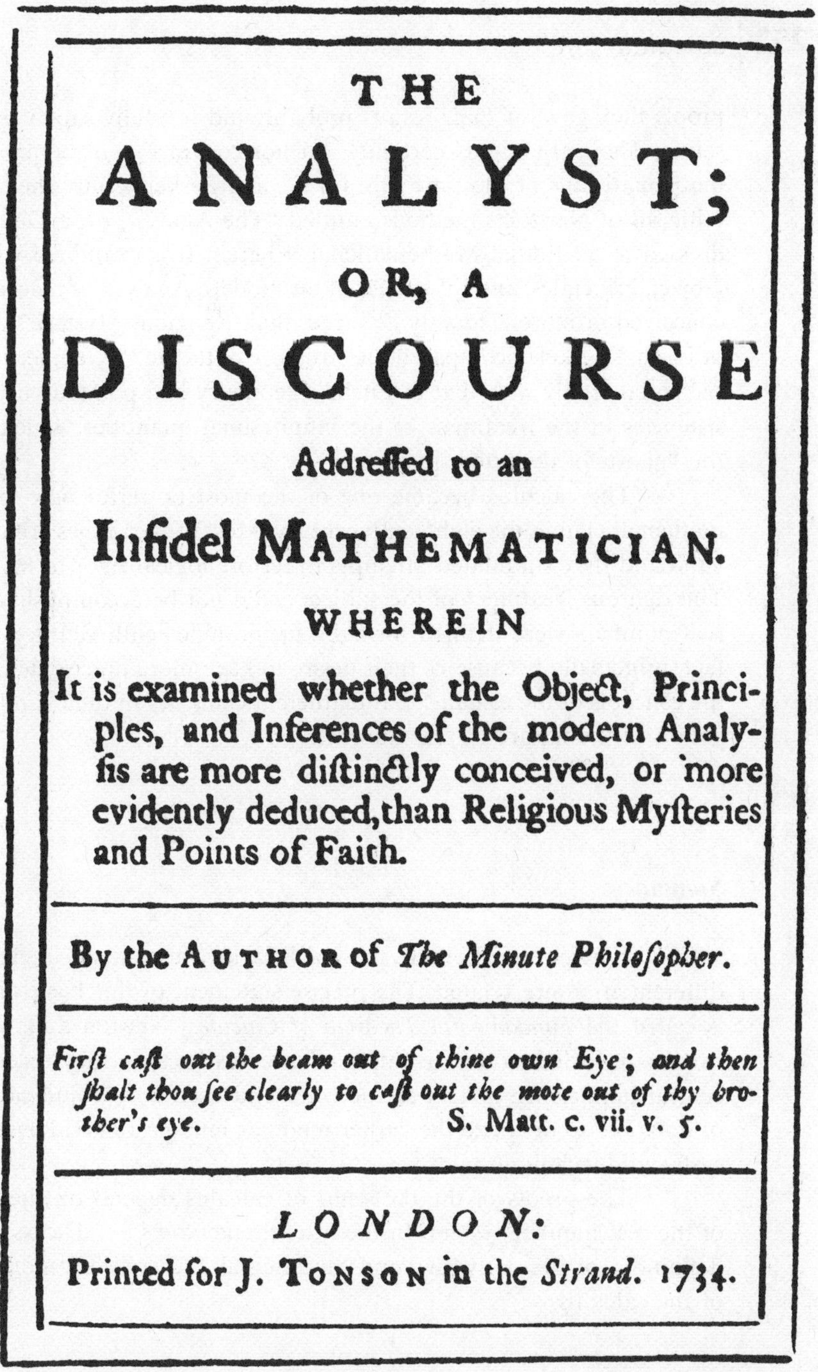
THE

ANALYST;

OR, A

DISCOURSE

Addressed to an

Infidel MATHEMATICIAN.

WHEREIN

It is examined whether the Object, Principles, and Inferences of the modern Analysis are more distinctly conceived, or more evidently deduced, than Religious Mysteries and Points of Faith.

By the AUTHOR of *The Minute Philosopher*.

First cast out the beam out of thine own Eye; and then shalt thou see clearly to cast out the mote out of thy brother's eye. S. Matt. c. vii. v. 5.

LONDON:

Printed for J. TONSON in the *Strand*. 1734.

THE ANALYST. *The title page from the volume written by Bishop Berkeley criticizing the logical foundations of calculus.* (*George Arents Research Library*)

proofs they gave of their results probably did not fully satisfy even themselves. Their arguments certainly did not convince many intelligent nonmathematicians of the time. Bishop George Berkeley published in 1734 a criticism of Newton's methods entitled "The Analyst; or, a Discourse Addressed to an Infidel Mathematician Wherein It is examined whether the Object, Principles, and Inferences of the modern Analysis are more distinctly conceived or more evidently deduced, than Religious Mysteries and Points of Faith." Berkeley compared the structure of the new "Geometrical Analysis" unfavorably with that of Euclid's geometry and pointed out the inconsistencies in the treatment of the infinitesimal quantities, which he called the "ghosts of departed quantities."

The calculus became one of the most powerful new tools of the mathematician in the eighteenth century. Most of Berkeley's criticisms were valid and they stimulated attempts to restore logical rigor to mathematics. The rigorous treatment of the subject could not be accomplished until the real numbers were defined formally in the nineteenth century. It was, in fact, principally because of their desire to have more precise definitions for the concepts of the calculus that mathematicians began their careful investigation of the nature of real numbers.

Summary

Newton and Leibniz showed how the methods of integration and differentiation are related. The precise statement of this basic relationship is called the *Fundamental Theorem of Calculus*. Newton and Leibniz are usually regarded as the inventors of calculus because they recognized the central importance of this relationship and used it, and judicious choices of notation, to organize the earlier methods into a set of widely applicable, systematic techniques.

The proofs of the theorems of calculus depend on the properties of the real number system. In the next chapter we shall discuss the formal definitions of the derivative and the integral and consider more theorems of the calculus.

EXERCISES

1. Use the Fundamental Theorem of Calculus to compute the integrals below, and verify the equations. Draw graphs to show the geometrical interpretation.

$$\int_0^2 3x = \int_0^1 3x + \int_1^2 3x \qquad \int_5^7 2x - 3 = \int_5^7 2x - \int_5^7 3$$

$$\int_1^4 x^2 + 6 = \int_1^4 x^2 + \int_1^4 6 \qquad \int_2^6 4x^3 = 4\int_2^6 x^3$$

2. If the velocity of an object at time t is given by the formula $v = t^3 - t^2 + 1$, find the distance traveled from time $t = 4$ to time $t = 5$. What is the formula for the acceleration of the object? Is it true that the velocity of the object is always increasing?

DISCUSSION QUESTION

Try to find and state general properties about the values of integrals. Use the examples of Exercise 1 as a starting point for your generalizations. Use the Fundamental Theorem of Calculus and geometric interpretations to justify your "theorems."

twenty-eight

analysis

The calculus is a powerful tool in mathematics and knowledge o its techniques is necessary for anyone engaged in serious scientific investi gations. In the last one hundred years mathematicians have greatly increase the usefulness of the calculus by extending the applicability of the methods The branch of mathematics that is involved with the study of the calculu and its many generalizations is called *analysis*. The derivative and th integral that have been considered in preceding chapters are properly par of *real analysis* since the domain of discussion was restricted to the set o real numbers. It is the purpose of this chapter to show how the definition of the calculus can be made precise in terms of the formally defined rea number system. Definitions like these are the starting point in the stud of analysis.

Limits

Augustin-Louis Cauchy (1789–1857) was one of the first men t contribute to the modern formulation of the calculus. Cauchy, a contem porary and countryman of Poncelet, graduated as a military engineer an helped design coastal defences for Napoleon against the English. He becam famous for his solutions to applied problems while quite young and wa one of the established mathematicians who ignored Abel and Galois. Cauch spent a few years in exile at the time of the revolution of 1830, but he hel important teaching or government positions most of his life.

Beginning about 1820, Cauchy worked on the foundations of ana ysis. His results influenced Cantor's definition of the real number system and his definitions for the calculus are essentially the same as those used today.

The basic concept in the calculus is that of a *limit*. The methods of differentiation and integration were both based on the idea of a number "becoming infinitesimally small" or one number "becoming very close" to another number "in the limit." Cauchy gave an arithmetic definition of this concept that freed it from dependence on geometric intuition and dynamic properties of numbers.

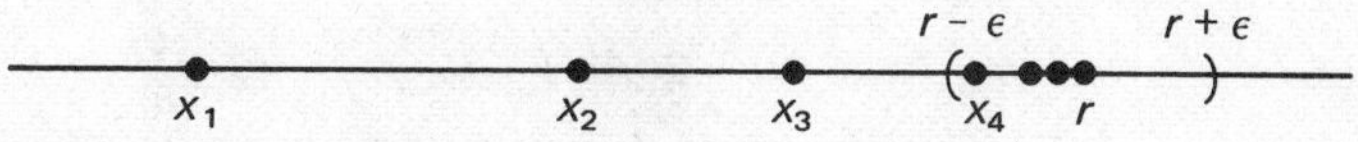

DEFINITION | A sequence $x_1, x_2, \ldots, x_n, \ldots$ of real numbers is said to **converge to the real number r** or **approach the limit r** if for every positive number ϵ there exists a natural number N such that

$$|x_n - r| < \epsilon$$

for every n greater than N.

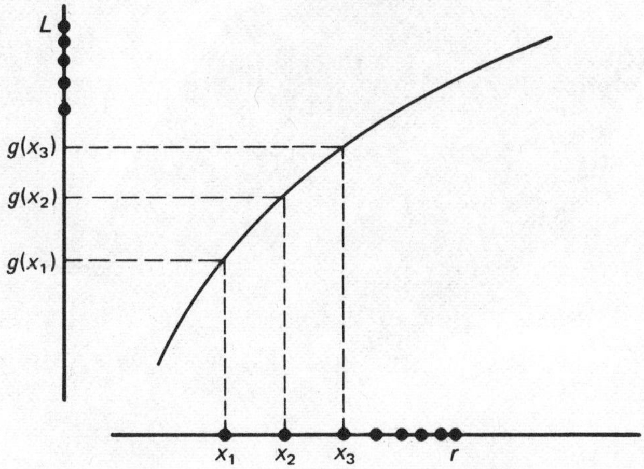

DEFINITION | Let g be a function defined on the real numbers. Then g **approaches the limit L as x approaches r** if for each sequence $x_1, x_2, \ldots$ of real numbers that converges to r, the corresponding sequence $g(x_1), g(x_2), \ldots$ converges to L. This will be denoted

$$\lim_{x \to r} g(x) = L.$$

Augustin-Louis Cauchy. *Cauchy was trained as a civil engineer and eventually became professor of mathematical astronomy at the Sorbonne. He was known for his ability to solve difficult practical problems.* (*George Arents Research Library*)

An intuitive understanding of these definitions is all that is necessary for our discussions. We shall not become involved with the computations necessary to prove, using the definitions, that sequences or functions approach limits. A numerical example may, however, help to clarify the definitions.

Example: The sequence $x_1 = 3$, $x_2 = 2\frac{1}{2}$, $x_3 = 2\frac{1}{3}$, $x_4 = 2\frac{1}{4}$, ... converges to the limit 2. This can be proved by showing that if *any* small number ϵ is given, then a value of n can be found such that all terms after the nth term are within ϵ of 2. (In this case, if $n > 1/\epsilon$, then $|x_n - 2| < \epsilon$.)

The function $f(x) = 3x$ approaches the limit 6 as x approaches 2. The function $f(x) = x^2 + 1$ approaches 10 as x approaches 3. These facts are restated below.

$$\lim_{x\to 2} 3x = 6 \qquad \lim_{x\to 3} (x^2 + 1) = 10$$

The condition of the definition of a convergent sequence of numbers is, essentially, that the sequence behave well enough to "define" some real number and that the real number "defined" be r. As a matter of fact, sequences of the type used in Chapter 16 to define real numbers are named in honor of Cauchy.

DEFINITION | A sequence $x_1, x_2, \ldots, x_n, \ldots$ of real numbers is said to be a **Cauchy sequence** if for every $\epsilon > 0$, there is a natural number N such that

$$|x_n - x_m| < \epsilon$$

for all n and m greater than N.

According to Cauchy, the most important property of real numbers is that sequences of this kind always converge to a real number. He tried in vain to prove this fact.

CAUCHY'S THEOREM | A sequence of real numbers converges to some real number if and only if it is a Cauchy sequence.

The statement of this theorem is equivalent to the completeness assumption about the real number system (see Axiom 14, p. 150). Cauchy could not prove the theorem because it is independent of the assumptions he made about numbers. The theorem is easily proved if the axioms of Chapter 16 or one of the definitions of real number given by Dedekind and Cantor is used.

The Derivative and the Integral

The modern definition of the derivative is simply a precise version of the one used by Fermat, Newton, and all the others. The derivative of a function at a point is defined to be the limit of a certain quotient function.

DEFINITION | Let f be a function defined on the interval $[a, b]$, and let x_1 be a point in the interval. Then f has a derivative at x_1 if

$$\lim_{x \to x_1} \frac{f(x) - f(x_1)}{x - x_1}$$

exists. If the limit exists, it is called the **derivative of f at x_1** and denoted $f'(x_1)$. The function f' is called the **derivative of f.**

Using this definition and the properties of the real numbers, it is possible to prove theorems about derivatives that are useful when one works with the calculus. For example, if f and g are suitable functions, formulas can be proved for finding the derivative of their sum, difference, product, and quotient.

$$(f + g)' = f' + g' \qquad (f \cdot g)' = f' \cdot g + f \cdot g'$$

$$(f - g)' = f' - g' \qquad \left(\frac{f}{g}\right)' = \frac{f'g - fg'}{g^2}$$

Example: Although we are not interested in developing computational skills, a simple application of these results should help clarify what they say. Let $f(x) = 4x + 3$ and $g(x) = x^2 + 1$. Then, $f'(x) = 4$ and $g'(x) = 2x$.

$$[f(x) + g(x)]' = (4x + 3)' + (x^2 + 1)' = 4 + 2x$$

$$[f(x) - g(x)]' = (4x + 3 - x^2 - 1)' = 4 - 2x$$

$$[f(x) \cdot g(x)]' = [(4x + 3)(x^2 + 1)]' = 4(x^2 + 1) + 2x(4x + 3)$$

$$\left[\frac{f(x)}{g(x)}\right]' = \left[\frac{4x + 3}{(x^2 + 1)}\right]' = \frac{4(x^2 + 1) - 2x(4x + 3)}{(x^2 + 1)^2}$$

The definition of the integral can be made precise in a similar fashion. The first step is to introduce notation for the sums to be used in the approximations. It is necessary to be very careful when the integral is formally defined as the limit of a sequence of sums. Technical details must be considered if the definition is to be precise. It is more convenient to use Dedekind's version of the definition of real numbers in this situation. For our purposes it will suffice to outline the procedure.

Let f be a function defined on an interval $[a, b]$. First, "lower sums" and "upper sums" are formed by subdividing the interval and adding up the areas of rectangles that are known to be too small and too large, respectively.

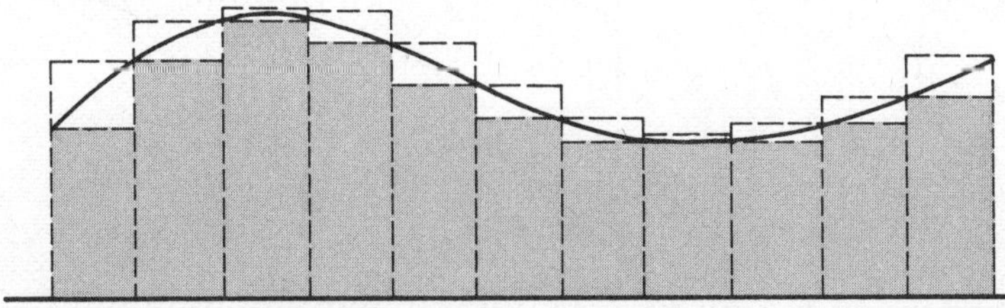

Then one considers the set of *all* numbers that can be obtained as lower sums when the interval is divided into subintervals *in any way*. This set of "lower approximations" is a set of real numbers that is bounded above. It therefore has a least upper bound s. Similarly, the set of numbers obtainable as upper sums is bounded below and has a greatest lower bound S.

If $S = s$, the integral

$$\int_a^b f$$

is said to exist and its value is the common value of S and s.

Whichever form of the definition of the integral is used, it is *very* difficult to apply the definition to compute integrals. It is possible to prove from the definition many theorems that can be used to simplify integration problems. Two of the most basic such facts are stated below.

$$\int_a^b f + \int_b^c f = \int_a^c f$$

$$\int_a^b (f + g) = \int_a^b f + \int_a^b g$$

Using results such as these and the Fundamental Theorem of Calculus, a great many integrals can be evaluated without having to use the definition of the integral directly. As a "last resort," of course, a program can be written instructing a machine to use the definition to provide an approximation to any desired degree of accuracy.

Analysis is not primarily the study of numerical results, however. It is the investigation of the methods of calculus, the situations in which they can be applied, and the properties of the results.

DISCUSSION QUESTIONS

1. Do you think that the function graphed below has a derivativ at every point of the interval?

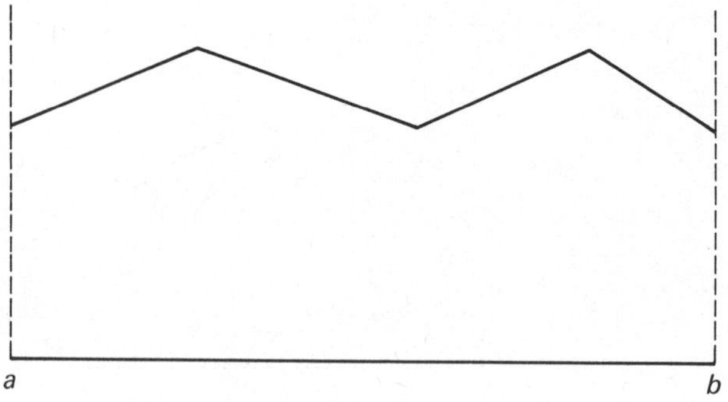

2. Do you think that the integral that would be denoted

$$\int_0^1 \frac{1}{x^2}$$

actually exists?

3. Check your intuitive ideas of what the definitions in this chapte: say by constructing some more examples and drawing some pictures. Shov on a real number line the locations of the points of sequences that do o: do not converge. Sketch the graphs of functions that have a derivative a some points but do not have a derivative at others. Try to sketch the grapł of a function that is not integrable.

4. Using the definition of this chapter, it can be shown that integrals exist for a wide variety of functions. It is natural to continue to interpre many integrals in terms of "area under the curve." Which of the followinǥ two statements do you think more accurately describes the situation?

(a) The integral can be used to evaluate the size of certain areas

(b) Area is (defined to be) the quantity obtained by integration

It may help to reconsider some simple cases such as the circle.

5. How do you think a quantity that would be denoted

$$\int_a^{+\infty} f$$

should be defined?

appendix

definitions and theorems from the elements

The following are the definitions and a selection from the 48 theorems of Book I of the *Elements* of Euclid.

DEFINITIONS

1. A *point* is that which has no part.
2. A *line* is length without breadth.
3. The extremities of a line are points.
4. A *straight line* is a line that lies evenly with the points of itself.
5. A *surface* is that which has only length and breadth.
6. The extremities of a surface are lines.
7. A *plane surface* is a surface that lies evenly between its extreme lines.
8. A *plane angle* is the inclination of two lines to one another if the lines meet and do not lie in a straight line.
9. A *rectilineal angle* is a plane angle in which the two lines are straight lines.
10. When a straight line standing on another straight line makes the adjacent angles equal to one another, each of the angles is called a *right angle,* and the straight line standing on the other is called a *perpendicular* to it.
11. An *obtuse angle* is an angle greater than a right angle.
12. An *acute angle* is an angle less than a right angle.
13. A *boundary* is that which is an extremity of anything.
14. A *figure* is that which is enclosed by any boundary or boundaries.
15. A *circle* is a plane figure contained by one line such that all straight lines drawn to it from a certain point within the figure are equal.
16. The point is called the *center* of the circle.

17. A *diameter* of a circle is a straight line drawn through the center o the circle and terminated at both ends by the circle, and a diameter bisect a circle.

18. A *semicircle* is the figure contained by a diameter and that part o the circumference cut off by the diameter.

19. *Rectilineal* figures are those that are contained by straight lines *trilateral* figures are those contained by three lines; *quadrilateral* are thos contained by four; and *multilateral* are those contained by more than fou lines.

20. Of the trilateral figures, an *equilateral triangle* is that which has thre equal sides; an *isosceles triangle* is that which has only two sides equal; anc a *scalene triangle* is that which has three unequal sides.

21. A *right-angled triangle* is that which has a right angle; an *obtuse angled triangle* is that which has an obtuse angle; and an *acute-anglec triangle* is that which has three acute angles.

22. Of the quadrilateral figures, a *square* is that which has all its sides equal and all its angles are right angles; an *oblong* has all right angles bu not all sides equal; a *rhombus* has all sides equal but its angles are not righ angles; and a *rhomboid* has opposite sides and angles equal but all sides are not equal and all angles are not right angles. All other quadrilaterals are called *trapezia.*

23. *Parallel* straight lines are straight lines that are in the same plane and, being produced indefinitely in both directions, do not meet.

THEOREMS

I-1. An equilateral triangle can be constructed on a given finite straight line.

I-2. A straight line equal to a given straight line can be drawn from a given point.

I-3. Given two unequal straight lines, a part equal to the lesser can be cut off the greater.

I-4. If two triangles have two sides equal to two sides, respectively, and have the angles contained by those sides equal to one another, then the third sides are equal and the areas are equal and the other angles will be equal, respectively, namely those opposite equal sides.

I-5. The angles at the base of an isosceles triangle are equal and if the equal sides are produced, then the angles on the other side of the base are also equal to each other.

I-6. If two angles of a triangle are equal to each other, then the sides opposite them are also equal.

I-8. If two triangles have two sides equal to two sides of the other, and the third sides are also equal, then the angle contained by the two sides of one is equal to the angle contained by the two sides of the other.

I-9. A given rectilineal angle can be bisected.

I-10. A given finite straight line can be bisected.

I-11. A perpendicular can be drawn to a given straight line at a given point.

I-12. A perpendicular can be drawn from a given point to a given infinite straight line.

I-13. The angles that one straight line makes with another, upon one side of it, are either two right angles or are together equal to two right angles.

I-15. If two straight lines intersect, then the vertical, or opposite, angles are equal.

I-16. If one side of a triangle is produced, then the exterior angle is greater than either of the interior and opposite angles.

I-17. Any two angles of a triangle are together less than two right angles.

I-18. In any triangle the greatest side is opposite the greatest angle.

I-19. In any triangle the greatest angle is opposite the greatest side.

I-20. Any two sides of a triangle are together greater than the third side.

I-22. A triangle can be constructed with sides equal to any three given straight lines provided that any two of these lines together is greater than the third.

I-23. A rectilineal angle equal to a given rectilineal angle can be constructed at a given point on a given straight line.

I-24. If two triangles have two sides equal to two sides, respectively, but the angle contained by one is greater than the other, then the third side will also be greater than the third side of the other.

I-25. If two triangles have two sides equal to two sides, respectively, but have the third side of one greater than the third side of the other, then the angle contained by the two sides of the one will be greater than that contained by the other two sides.

I-26. If two triangles have two angles equal to two angles, respectively, and one side equal to one side, either the side adjacent to the equal angles or a side opposite one of the equal angles, then the remaining sides are equal, respectively, and the third angle of one equals the third angle of the other.

I-27. If a straight line falling upon two other straight lines makes the alternate angles equal, then the straight lines are parallel.

I-28. If a straight line falling on two straight lines makes the exterior angle equal to the interior and opposite angle on the same side, or the interior angles on the same side equal to two right angles, then the two straight lines are parallel.

I-29. If a straight line falls on two parallel straight lines, it makes the alternate angles equal to each other, the exterior angle equal to the opposite and interior angle, and the interior angles on the same side equal to two right angles.

I-30. Straight lines parallel to the same straight line are parallel to each other.

I-31. Through a given point a straight line parallel to a given straight line can be constructed.

I-32. If the side of a triangle is produced, then the exterior angle is equal to the two interior and opposite angles and the three interior angles are equal to two right angles.

I-33. The straight lines that join the extremities of two equal and parallel straight lines are themselves equal and parallel.

I-46. A square can be constructed on a given straight line.

I-47. In a right-angle triangle the square of the side opposite the right angle is equal to the squares of the sides containing the right angle.

I-48. If the square of one of the sides of a triangle is equal to the squares of the other two sides, then the angle contained by the two sides is a right angle.

selected bibliography

Aside from the inclusion of some of the works actually mentioned in this book, the following sources were selected because they are readable and generally available. Those works marked with an asterisk contain extensive bibliographies for the history and development of mathematics.

ADLER, C. F., *Modern Geometry, an Integrated First Course*. New York: McGraw-Hill Book Co., 1958.

ALBERTI, L. B., *On Painting,* trans. by John Spencer. New Haven: Yale University Press, 1966.

BARROW, ISAAC, *Lectiones Opticae et Geometricae*. London, 1674.

BELL, E. T., *Men of Mathematics*. New York: Simon & Schuster, Inc., 1937.

BLUMENTHAL, L. M., *A Modern View of Geometry*. San Francisco: W. H. Freeman and Co., 1961.

BONOLA, ROBERTO, *Non-Euclidean Geometry: A Critical and Historical Study of Its Developments*. New York: Dover Publications, Inc., 1955.

*BOYER, CARL B., *A History of Mathematics*. New York: John Wiley & Sons, Inc., 1968.

CARDANO, J., *The Book of My Life,* trans. by J. Stoner. New York: Dover Publications, Inc., 1962.

CHACE, A. B., L. S. BULL, H. P. MANNING, and R. C. ARCHIBALD, *The Rhind Mathematical Papyrus*. Oberlin, Ohio: M.A.A., 1927–1929, 2 vol.

DESCARTES, RENÉ, *The Geometry of René Descartes,* trans. by D. E. Smith and M. L. Latham. Chicago: Open Court Publishing Co., 1925.

EVANS, J. P., *Mathematics: Creation and Study of Form*. Reading, Mass.: Addison-Wesley Publishing Co., 1970.

*EVES, HOWARD, *An Introduction to the History of Mathematics.* New York: Rinehart & Company, Inc., 1953.

EVES, HOWARD, *A Survey of Geometry, Vol. I.* Boston: Allyn and Bacon, Inc., 1963.

FRALEIGH, J. B., *Mainstreams of Mathematics.* Reading, Mass.: Addison-Wesley Publishing Co., 1969.

GAUSS, C. F., *Disquisitiones Arithmeticae,* trans. by A. A. Clarke. New Haven: Yale University Press, 1966.

HALL, TORD, *Carl Friedrich Gauss,* trans. by Albert Froderberg. Cambridge, Mass.: The M.I.T. Press, 1971.

HAWKINS, GERALD S., *Stonehenge Decoded.* New York: Dell Publishing Co., 1965.

HEATH, T. L., *A History of Greek Mathematics.* Oxford: The Clarendon Press, 1921, 2 vol.

HEATH, T. L., *The Thirteen Books of Euclid's 'Elements,'* 2nd ed. Cambridge: The University Press, 1926, 3 vol.

INFELD, L., *Whom the Gods Love.* New York: Whittlesey House, 1948.

KLINE, MORRIS, ed., *Mathematics in the Modern World.* New York: W. H. Freeman and Co., 1968.

LANDAU, E., *Foundations of Analysis,* trans. by S. Steinhardt. New York: Chelsea Publishing Co., 1951.

LEVY, MARTIN, *The 'Algebra'* of Abū Kāmil. Madison: University of Wisconsin Press, 1966.

NEUGEBAUER, O., and A. SACHS, *Mathematical Cuneiform Texts.* New Haven, Conn.: American Oriental Society, 1945.

ORE, OYSTEIN, *Cardano, The Gambling Scholar.* Princeton: Princeton University Press, 1953.

ORE, OYSTEIN, *Niels Henrik Abel, Mathematician Extraordinary.* Minneapolis: University of Minnesota Press, 1957.

PEANO, G., *Formulario Mathematico.* Rome: Edizioni Cremonese, 1960.

PLAYFAIR, JOHN, *Elements of Geometry.* Philadelphia: Marot and Walter, 1825.

SACCHERI, GIROLAMO, *Euclides Vindicatus,* ed. and trans. by G. B. Halsted. Chicago: Open Court Publishing Co., 1920.

*SMITH, D. E., *History of Mathematics.* Boston: Ginn and Company, 1923, 2 vol.

*STRUIK, D. J., *A Concise History of Mathematics.* New York: Dover Publications, Inc., 1948, 2 vol.

STRUIK, D. J., *A Source Book in Mathematics, 1200–1800.* Cambridge: Harvard University Press, 1969.

TIETZE, H., *Famous Problems of Mathematics.* Baltimore: Graylock Press, 1965.

VAN DER WAERDEN, B. L., *Science Awakening,* trans. by Arnold Dresden. New York: Oxford University Press, 1961.

ndex